MATH HOMEWORK ASSIGNMENTS

GRADE 5

by Jenni Harrold

World Teachers Press

Published with the permission of R.I.C. Publications Pty. Ltd.

First published by R.I.C. Publications Pty. Ltd., Perth, Western Australia.

Printed in the United States of America.

Order Number 2-5075
ISBN 1-885111-91-6

A B C D E F 99 00 01 02

Didax
Educational Resources

395 Main Street
Rowley, MA 01969

Foreword and Contents

Math Homework Assignments is an excellent workbook which covers all components in the major curriculum areas of number, space, and measurement. Each assignment focuses on one concept and the questions show a varying degree of difficulty. The assignments can be used for homework, to reinforce class work, and as a review during class sessions. All the concepts are clearly defined and sequentially ordered.

Contents

Teacher Information – *Math Homework Assignments*

Introduction

Consolidation and review activities are recognized as an integral part of learning and understanding a specific concept. Once a concept has been taught, children need several opportunities to practice, develop and understand the ideas and methods behind it.

Math Homework Assignments provides exactly this opportunity. The reinforcement of concepts through an assignment approach allows you to readily evaluate where each child may be having difficulties and provides parents with the opportunity to see how their child is achieving in mathematics.

The thirty-two assignments and four review sheets follow this format, allowing the child to focus solely on the task at hand. This also develops a routine which aids the learning process and allows the child to attack independent activities/homework with confidence.

Strategies

Math Homework Assignments can be used as:

1. Homework activities

   Each assignment can be photocopied and sent home for children to complete independently over the course of the week or a pre-designated time period. Parents/guardians can assist the child if they are having difficulties. The following approaches are encouraged:

   (a) Assist the child with the process involved without solving the problem for the child, and

   (b) encourage the child to try and solve the problem.

   Any problems encountered at home should be discussed with you at the earliest possible convenience. Each assignment focuses on one concept and the questions show a varying degree of difficulty. For this reason, the assignments can be successfully used for:

2. Consolidation of class work

   Once a concept has been presented in class, the appropriate assignment can be photocopied and completed in class to consolidate and reinforce the concept.

3. Review of class work

   At the end of a unit of work, the assignments can be used to assess the child's understanding of a particular concept. This allows you to focus further instruction at the point of need for individual children. This approach also provides a straightforward approach to evaluation and recording of the child's mathematical ability.

Instructions

The instructions provided on each assignment are clear and concise. Each instruction has been carefully written to avoid any ambiguity. This allows children to work as independently as they have no need to clarify the question.

Benefits

The benefits of *Math Homework Assignments* are many.

1. You can readily evaluate where each child is having success or difficulties.
2. Provides parents with the opportunity to observe how their child is achieving in the area of mathematics.
3. Opens communication between school and home regarding each child's progress in mathematics.
4. Opportunities are provided for children to practice, consolidate and review various concepts treated in class.
5. Children are able to take some responsibility for their own learning.

Conclusion

Math Homework Assignments is a useful tool for developing the knowledge and understanding of a broad range of mathematical concepts. Children can develop a high level of confidence with the opportunities they are given to consolidate what they learn. Confidence in children leads to success and a positive self-image of themselves as learners.

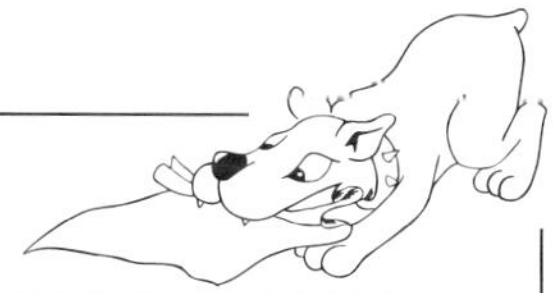

Assignment 1 – Addition

1.

4 + 8 = ____	9 + 5 = ____	7 + 9 = ____	4 + 12 = ____
6 + 6 = ____	5 + 13 = ____	12 + 5 = ____	14 + 9 = ____
13 + 8 = ____	5 + 14 = ____	25 = 12 = ____	12 + 9 = ____
20 + 9 = ____	22 + 8 = ____	15 + 15 = ____	14 + 21 = ____

2.

26	56	41	51	74	48
+ 22	+ 32	+ 37	+ 48	+ 26	+ 32
____	____	____	____	____	____

123	204	342	286	555	199
+ 243	+ 245	+ 139	+ 399	+ 309	+ 286
____	____	____	____	____	____

3. *Danny has 48 trading cards, Craig has 72, Frank has 101 and Julie has 56. How many trading cards are there altogether?* ____________

4. *A small movie theater recorded the following attendance figures during one week – 209, 145, 186, 223, 309, 358, 238. How many people attended altogether?* ____________

5. *For a fundraiser students in a school collected the following amounts of money – Grade 1 – $138, Grade 2 – $98, Grade 3 – $94, Grade 4 – $109, Grade 5 – $136, Grade 6 – $229 and Grade 7 – $258. How much money was collected altogether?* ________________

6. *Two International Soccer Matches were played over a weekend. If 46,780 people attended on Saturday and 54,928 people attended on Sunday what was the total attendance?* ________________

7. *At an auction, five cars were sold for the following amounts – $6,000; $8,450; $8,990; $12,500; and $10,555. What was the total amount raised?* __________

8. *Three concerts were held during the week. If 7,780 people attended the first night, 6,975 the second, and 7,648 the third, how many attended altogether?* ____________

9. *After a lottery win, a family purchased a new television set for $529, a video recorder for $459, a microwave for $390, a portable stereo for $679, and a couch for $899. How much did they spend altogether?* ________________

10. *For the school council elections, Sandy received 58 votes, Tammy received 39 votes, Justin received 22 votes, Sean received 47 votes and, Wendy received 64 votes. How many votes were taken altogether?* ______________

11. *During a ratings survey, the top three television programs attracted 2,428,000; 2,201,050; and 1,908,000 viewers. How many viewers watched these programs?* ____________

Assignment 2 – Subtraction

1. 59 − 42 = ______

2. 84 − 43 = ______

3. 356 − 142 = ______

4. 256 − 150 = ______

5. 707 − 407 = ______

6. 999 − 606 = ______

7. 42 − 25 = ______

8. 51 − 38 = ______

9. 242 − 139 = ______

10. 352 − 178 = ______

11. 508 − 229 = ______

12. 3,456 − 1,789 = ______

13. *Subtract $23.55 from $42.65.* ______

14. *Subtract 125,545 from 150,000.* ______

15. *The book I am reading has 407 pages. If I am up to page 278, how many more pages do I need to read before the end?* ______

16. *Amanda has saved $8,275 towards a deposit for a new apartment. If she needs $12,000, how much more does she need to save?* ______

17. *In a census there were 20,635 people living in a town. Five years later there were 24,798 people. By how much did the population increase?* ______

18. *A fish tank held 20 liters of water. If there was already 13.5 liters in the tank, how much more water was needed to fill it?* ______

19. *2,106 people attended a show on Saturday and 6,879 people attended on Sunday. What was the difference in attendance?* ______

20. *Mr. and Mrs. Jones sold their house for $146,000. If they bought it for $119,750, how much profit did they make?* ______

Girls	
Jane	14.5 m
Anne	17.8 m
Jenny	16.4 m
Cathy	17.3 m
Lisa	14.9 m

Boys	
Peter	15.9 m
Rick	22.7 m
Scott	18.6 m
Jeff	20.2 m
Mark	21.8 m

At the school carnival the following distances were recorded for the discus throw.

21. *What is the difference between the longest girls' and boys' throw?* ______

22. *What is the difference between the longest and shortest girls' throw?* ______

23. *What is the difference between the longest and shortest boys' throw?* ______

24. *What is the difference between Jane's throw and Rick's throw?* ______

Assignment 3 – Multiplication

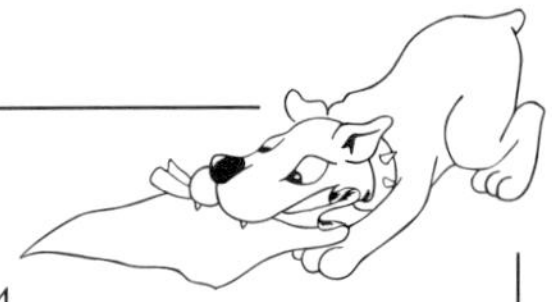

1.

23	41	53	36	55	84
× 2	× 2	× 3	× 4	× 6	× 4

2.

122	134	116	205	319	525
× 2	× 2	× 5	× 6	× 4	× 5

3.

50	20	38	78	67	99
× 21	× 32	× 25	× 40	× 50	× 30
+	+	+	+	+	+

4.

124	222	308	400	509	918
× 12	× 40	× 32	× 75	× 55	× 50
+	+	+	+	+	+

5. *Sarah bought six books that were $2.50 each. How much did she pay altogether?* ______

6. *A group of 20 children each paid $4.50 to see their school play. How much did it cost altogether?* ______

7. *Each week Sean traveled 37 km on his bike. How far would he ride in 7 weeks?* ______

8. *A small pool held 1,248 liters of water. How much water would the large pool hold if it was eight times bigger?* ______

9. *If one carton holds 12 eggs, how many eggs will there be in 36 cartons?* ______

10. *If a plane carries 389 passengers on each flight, how many passengers would it carry on 32 flights?* ______

11. *If there were 188 children in each of the four grades, how many children were there altogether?* ______

12. *A group of 78 children each paid $2.20 for a field trip. What was the total cost?* ______

13. *Sam delivered 379 leaflets each week. How many would he deliver if he maintained this number for 15 weeks?* ______

14. *Complete the grids.*

6					
5					
4					
3					
2					
x	2	3	4	5	6

11					
10					
9					
8					
7					
x	11	10	9	8	7

Assignment 4 – Division

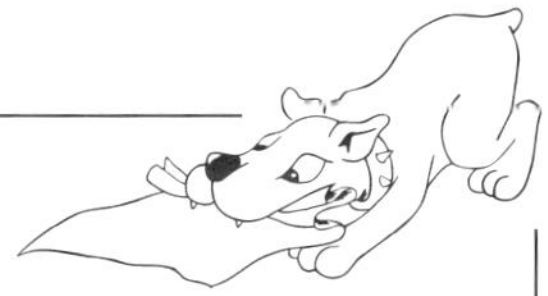

1. *Complete the following.*

6) 36   5) 40   7) 49   2) 48   3) 96   4) 88

6) 126   4) 124   5) 155   3) 186   7) 217   6) 186

2. *Divide $30 among 5 people.* ________   *404 books among 4 students.* ________

3. *Share $66 among 6 children.* ________   *$393 among 3 people.* ________

4. *Complete the following.*

4) 56   3) 42   4) 92   2) 36   3) 87   4) 68

3) 426   5) 405   4) 564   4) 364   5) 705   6) 786

5. *A piece of wood is 405 cm long. How many 5 cm pieces can be cut from the wood?* ________

6. *Six buses took 270 children to the theater. How many children were on each bus?* ________

7. *If 900 hectares were divided into 6 equal sized fields, how many hectares were in each field?* ________

8. *How many teams of 8 could be made from 120 children?* ________

9. *Complete the following.*

4) 415   6) 615   7) 736   3) 803   7) 519   6) 737

4) 922   6) 858   5) 398   9) 593   3) 875   10) 349

10. *7 trucks carried 966 cartons. How many cartons did each truck carry?* ________

11. *I took $686 on a seven-day vacation. If I divided my money equally over the week, how much did I spend each day?* ________

12. *Complete the crossword puzzle.*

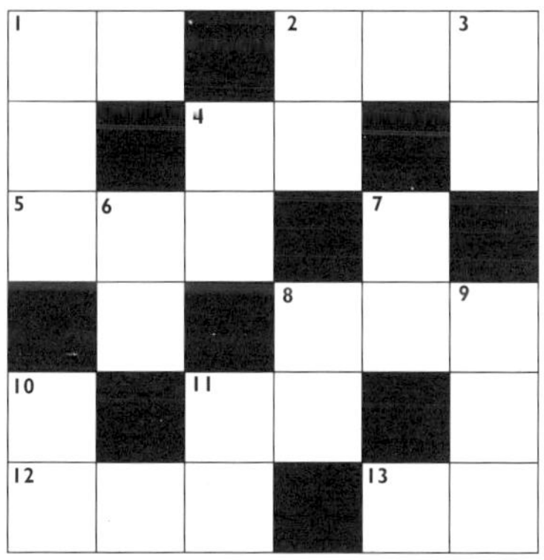

Across		Down	
1	189 ÷ 9	1.	924 ÷ 4
2.	978 ÷ 3	2.	266 ÷ 7
4.	196 ÷ 7	3.	320 ÷ 5
5.	600 ÷ 5	4.	160 ÷ 8
8.	847 ÷ 7	6.	162 ÷ 6
11.	456 ÷ 8	7.	288 ÷ 9
12.	840 ÷ 6	8.	68 ÷ 4
13.	147 ÷ 3	9.	447 ÷ 3
		10.	426 ÷ 6
		11.	400 ÷ 8

Assignment 5 – Mixed Problems

A	B	C	D	E	F
7 + 2 =	6 x 3 =	8 ÷ 8 =	9 – 1 =	8 x 4 =	7 x 7 =
12 ÷ 3 =	9 + 8 =	6 x 4 =	15 + 7 =	15 ÷ 5 =	20 – 9 =
8 x 5 =	20 ÷ 4 =	15 + 12 =	6 x 6 =	20 + 7 =	15 ÷ 15 =
22 – 9 =	17 – 7 =	15 – 9 =	21 ÷ 7 =	19 – 8 =	31 + 9 =
18 + 6 =	5 x 8 =	24 ÷ 6 =	22 – 12 =	7 x 8 =	3 x 6 =
18 ÷ 6 =	13 + 13 =	2 x 9 =	11 + 1 =	12 ÷ 2 =	13 – 8 =
9 x 0 =	8 ÷ 2 =	20 + 10 =	5 x 5 =	9 + 9 =	20 ÷ 5 =
20 – 10 =	9 – 9 =	18 – 0 =	30 ÷ 5 =	17 – 0 =	12 + 7 =
5 + 5 =	1 + 2 =	10 ÷ 5 =	12 – 11 =	5 x 10 =	11 x 2 =
9 ÷ 3 =	7 + 12 =	8 x 3 =	20 + 20 =	20 ÷ 10 =	16 – 8 =
7 x 4 =	2 ÷ 2 =	13 + 8 =	12 x 2 =	9 + 10 =	12 ÷ 1 =
10 – 8 =	11 – 7 =	15 – 5 =	14 ÷ 7 =	20 – 2 =	10 + 5 =
2 + 12 =	4 x 9 =	22 ÷ 11 =	21 – 1 =	2 x 12 =	12 x 3 =
15 ÷ 3 =	8 + 8 =	9 x 2 =	23 + 2 =	6 ÷ 2 =	15 – 3 =
6 x 5 =	55 ÷ 5 =	9 + 20 =	3 x 8 =	11 + 11 =	20 ÷ 2 =
20 – 5 =	18 – 4 =	12 – 0 =	18 ÷ 2 =	10 – 9 =	14 + 7 =
10 + 0 =	3 x 11 =	30 ÷ 3 =	20 – 1 =	11 x 4 =	5 x 7 =
21 ÷ 3 =	11 + 12 =	3 x 3 =	17 + 3 =	24 ÷ 1 =	12 – 8 =
4 x 4 =	16 ÷ 4 =	8 + 0 =	8 x 0 =	2 + 16 =	16 ÷ 8 =
20 – 15 =	9 – 3 =	15 – 2 =	50 ÷ 5 =	50 – 20 =	20 + 9 =
6 + 6 =	5 x 7 =	40 ÷ 5 =	31 – 1 =	2 x 2 =	8 x 8 =
25 ÷ 5 =	100 + 1 =	6 x 0 =	50 + 0 =	30 ÷ 2 =	22 – 3 =
10 x 4 =	11x 0 =	9 + 12 =	5 x 4 =	6 + 16 =	35 ÷ 7 =
9 – 7 =	13 – 3 =	15 – 0 =	48 ÷ 6 =	11 – 1 =	21 + 7 =

Assignment 6 – Rounding

0 5 10 15 20 25 30 35 40 45 50 55 60 65 70 75 80 85 90 95 100

1. *Round these numbers to the nearest 10.*

21 ______ 46 ______ 8 ______ 16 ______ 98 ______

33 ______ 75 ______ 88 ______ 92 ______ 447 ______

2. *Round these numbers to the nearest 100.*

111 ______ 623 ______ 299 ______ 450 ______ 666 ______

902 ______ 428 ______ 671 ______ 505 ______ 2,890 ______

3. *Round these numbers to the nearest 1,000.*

1,699 ______ 4,500 ______ 3,801 ______ 6,709 ______

9,005 ______ 5,500 ______ 12,050 ______ 23,690 ______

4. *Round the following times to the nearest minute.*

2 min 8 s ______ 15 min 40 s ______ 10 min 10 s ______ 59 min 1 s ______

16 min 4 s ______ 30 min 30 s ______ 45 min 45 s ______ 14 min 21 s ______

5. *Round these numbers to the nearest 10 and add them to find the approximate answer.*

24, 98, 12, 52, 69 ______

11, 92, 97, 22, 101 ______

109, 142, 151, 126, 198 ______

6. *Estimate these answers by rounding to the nearest 10.*

78 + 29 ~ 80 + 30 = 110 32 + 38 ~ ______

56 + 72 ~ ______ 91 + 97 ~ ______

79 – 12 ~ ______ 81 – 58 ~ ______

62 – 38 ~ ______ 55 – 27 ~ ______

41 × 22 ~ ______ 78 × 11 ~ ______

54 × 59 ~ ______ 83 × 48 ~ ______

7. *Round each number to the nearest 10 and then find the approximate answer.*

229 – 51 ~ ______ 657 – 248 ~ ______

83 × 12 ~ ______ 78 × 6 ~ ______

368 – 179 ~ ______ 7,342 + 7,143 ~ ______

264 + 791 ~ ______ 4,261 + 8,736 ~ ______

8. *Use your calculator to work out the exact answers to the above problems.*

Assignment 7 – Place Value

1. *Write the following numbers using digits.*

thirty-two = ____________

two thousand and six = ____________

one hundred and seven = ____________

six hundred and fifty-five = ____________

eighty-six thousand and five = ____________

three hundred and twenty-one = ____________

one thousand four hundred and fifty-seven = ____________

twenty thousand five hundred and seventy-three = ____________

2. *Write the place value and meaning of the underlined digits.*

	Place Value	**Meaning**
5<u>7</u>		
<u>2</u>7		
<u>1</u>59		
30<u>8</u>		
<u>5</u>91		
1,3<u>6</u>7		
<u>2</u>,907		
5,00<u>1</u>		

3. *Write the value of the 8 in each of the following:*

28 = ____________ 184 = ____________ 789 = ____________

803 = ____________ 8,924 = ____________ 18,675 = ____________

10,008 = ____________ 80,000 = ____________ 14,892 = ____________

4. *Write the largest value 5 digit number using:*

1, 3, 5, 7, 9 = ____________ 0, 2, 4, 6, 8 = ____________

7, 6, 5, 7, 6 = ____________ 9, 1, 0, 6, 8 = ____________

5. *Write the smallest value 5 digit number using:*

6, 0, 3, 9, 8 = ____________ 4, 9, 2, 7, 9 = ____________

2, 3, 5, 6, 0 = ____________ 9, 8, 7, 1, 4 = ____________

6. *Complete the following:*

2,534 = ____*thousands,* ____ *hundreds,* ____*tens,* ____*ones*

36,491 = ____*thousands,* ____ *hundreds,* ____*tens,* ____*ones*

15,679 = ____*thousands,* ____ *hundreds,* ____*tens,* ____*ones*

5,000 = ____*thousands,* ____ *hundreds,* ____*tens,* ____*ones*

84,037 = ____*thousands,* ____ *hundreds,* ____*tens,* ____*ones*

70,500 = ____*thousands,* ____ *hundreds,* ____*tens,* ____*ones*

Assignment 8 – Number Sentences

1. *Complete the following number sentences.*

6 + 9 = _____ 32 ÷ 4 = _____ 17 – 9 = _____ 23 + _____ – 45

72 – _____ = 47 7 × _____ = 56 7 × 4 + _____ = 36 22 + _____ + _____ = 36

2. *Solve the following by working out the problem in brackets first.*

(3 + 5) × 2 = _______ 50 ÷ (5 + 5) = _______ (29 – 8) ÷ 3 = _______

4 × (25 – 5) = _______ 18 + (4 × 5) = _______ (18 – 9) ÷ (36 – 33) = _______

3. *Add <, >, or = to make these true.*

$\frac{3}{4}$ ____ $\frac{1}{2}$ 2.5 ____ 5.5 6 + 3 ____ 18 – 8

4 × 8 ____ 40 – 2 1 week ____ 9 days 1,500 g ____ 1 kg

25 + 50 ____ 110 – 10 2.5 cm ____ 2.5 mm 7 × 8 ____ 50 + 6

4. *Write true or false for each of these.*

2 + 14 – 6 > 5 × 3 __________ (9 + 6) ÷ 3 < 8 – 4 + 3 __________

16 × (4 + 6) = 150 __________ (11 × 3) + 3 > 7 × 8 __________

(44 – 4) ÷ (8 – 6) < 40 __________ (6 + 5) × 2 = (5 + 6) × 2 __________

5. *Complete these number sequences.*

2, 4, 6, 8, ____, ____, ____, ____ 2, 4, 8, 16, ____, ____, ____, ____

5, 10, ____, ____, ____, 30 ____, ____, ____, ____, 50, 60, 70

3, ____, ____, ____, 15, ____ 5, 8, 11, ____, ____, ____, ____

6. *Complete one side of these number sentences to make them true.*

__________ = 20 __________ < 32 __________ > 50

7 × 8 < _____ 8 + 2 + 6 > _____ 100 = __________

_____ + _____ > 75 _____ < (4 × 3) + 12 7 + 7 + 7 < _____

7. *Complete the number grids.*

20					
19					
18					
17					
16					
–	5	7	9	12	14

9					
8					
7					
6					
5					
+	10	11	12	13	14

50					
40					
30					
20					
10					
+	11	22	33	44	55

100					
80					
60					
40					
20					
–	4	7	12	15	19

Assignment 9 – Multiplication

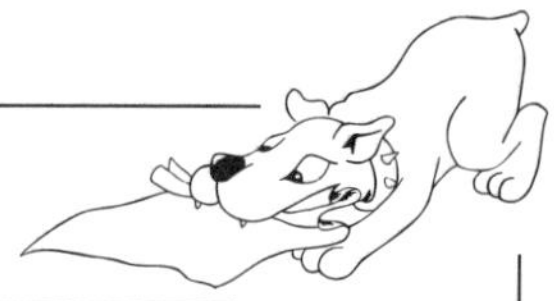

2 × 3 =	4 × 8 =	11 × 2 =	5 × 5 =	6 × 3 =
5 × 2 =	3 × 4 =	3 × 6 =	2 × 7 =	3 × 3 =
2 × 9 =	5 × 10 =	2 × 8 =	6 × 2 =	5 × 5 =
2 × 10 =	2 × 4 =	8 × 5 =	6 × 4 =	3 × 11 =
5 × 3 =	4 × 6 =	2 × 12 =	4 × 11 =	8 × 2 =
5 × 4 =	8 × 3 =	4 × 10 =	3 × 9 =	10 × 2 =
10 × 5 =	6 × 6 =	2 × 11 =	7 × 4 =	5 × 8 =
4 × 9 =	4 × 7 =	9 × 3 =	3 × 5 =	7 × 2 =
3 × 8 =	7 × 6 =	4 × 5 =	3 × 12 =	2 × 2 =
5 × 7 =	5 × 6 =	3 × 10 =	5 × 9 =	3 × 7 =
4 × 3 =	5 × 11 =	2 × 6 =	7 × 3 =	4 × 12 =
2 × 5 =	4 × 2 =	6 × 8 =	9 × 2 =	12 × 3 =
6 × 7 =	7 × 5 =	5 × 12 =	3 × 2 =	10 × 3 =
8 × 4 =	6 × 5 =	8 × 6 =	6 × 9 =	11 × 3 =
12 × 2 =	0 × 8 =	10 × 4 =	6 × 10 =	9 × 4 =
11 × 5 =	7 × 7 =	6 × 11 =	9 × 5 =	6 × 12 =
0 × 4 =	11 × 4 =	9 × 0 =	10 × 6 =	8 × 7 =
12 × 4 =	7 × 9 =	11 × 6 =	7 × 11 =	8 × 8 =
8 × 10 =	12 × 5 =	8 × 11 =	0 × 11 =	9 × 7 =
7 × 12 –	8 × 9 =	10 × 7 =	10 × 10 =	11 × 8 =
10 × 11 =	9 × 9 =	11 × 10 =	8 × 12 =	9 × 8 =
11 × 7 =	9 × 11 =	10 × 8 =	11 × 10 =	9 × 12 =
12 × 8 =	10 × 9 =	11 × 11 =	10 × 12 =	12 × 7 =
11 × 9 =	11 × 12 =	12 × 9 =	12 × 10 =	12 × 1 =
12 × 12 =	9 × 6 =	7 × 10 =	12 × 6 =	7 × 8 =

Assignment 10 – Decimals

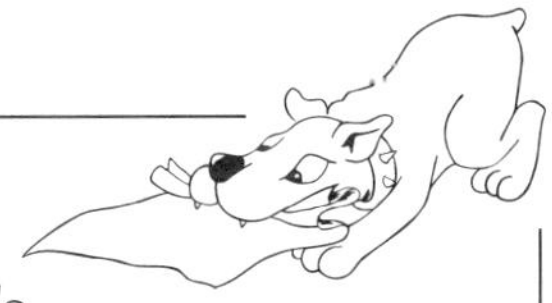

1. *Make a list of the different places and times where you might see or use decimals.*

____

____

2. *Put these decimals in order from smallest to largest.*

(a) 2.4, 2.8, 1.0, 1.90, 3.05, 3.50 ____________

(b) 4.88, 4.08, 8.48, 8.08, 4.00, 8.88 ____________

(c) 21.5, 12.5, 25.1, 25.5, 12.05, 12.55 ____________

(d) 2.99, 32.09, 32.90, 32.0, 32.32, 32.02 ____________

3. *Use < or > to make the following true.*

2.5 ____ 3.0	1.0 ____ 1.5	1.4 ____ 2.0	3.6 ____ 6.3
8.9 ____ 9.8	2.25 ____ 2.55	6.16 ____ 61.6	11.1 ____ 1.11
0.55 ____ 1.55	4.40 ____ 0.44	7.77 ____ 0.777	15.51 ____ 1.551

4. *Round these decimals to the nearest whole number.*

2.6 ____	8.9 ____	12.2 ____	45.6 ____
23.4 ____	26.0 ____	37.3 ____	58.8 ____
4.78 ____	9.01 ____	5.04 ____	28.89 ____

5. *In a long jump event six children recorded the following first round jumps –* 3.94 m, 2.87 m, 3.56 m, 3.96 m, 3.84 m, *and* 3.04 m. *Place the jumps in order from shortest to longest.* ____________

6. *Add these sets of decimals. Be sure to put them in correct columns.*

(a) 72.6 + 33.5 + 60.4 + 85.7= ____________

(b) 7.95 + 67 + 61.5 + 0.9 + 8.3 = ____________

(c) 5.3 + 0.07 + 7.8 + 2.47 + 10 = ____________

(d) 32 + 0.5 + 65.9 + 9 + 6.03 + 4.3 = ____________

(e) 84 + 9.7 + 0.8 + 65.9 + 42 + 1.456 = ____________

7. *Four children paid* $3.50 *to attend a play. How much did it cost altogether?* ____________

8. *Add the following weights –* 42.5 kg, 39.8 kg, 40 kg, 41.2 kg, and 42 kg. ____________

9. *Mary had* $567.25 *in one account and* $198.98 *in another. How much altogether?* ____________

10. *Tony is saving to buy a new bike. In the first month he saved* $42.50, *in the second month he saved* $38.90, *in the third month he saved* $55.65, *and in the fourth month he saved* $29. *How much did he save altogether?* ____________
If the bike he wanted was $227.55 *how much did he still need to save?* ____________

Assignment 11 – Special Numbers

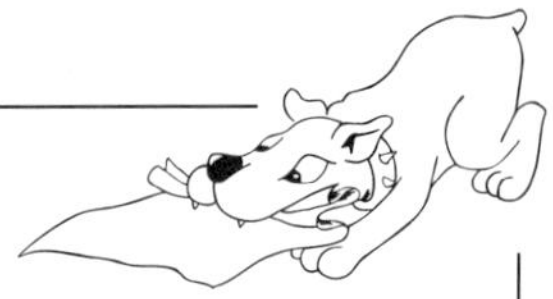

1	2	3	4	5	6	7	8	9	10
11	12	13	14	15	16	17	18	19	20
21	22	23	24	25	26	27	28	29	30
31	32	33	34	35	36	37	38	39	40
41	42	43	44	45	46	47	48	49	50
51	52	53	54	55	56	57	58	59	60
61	62	63	64	65	66	67	68	69	70
71	72	73	74	75	76	77	78	79	80
81	82	83	84	85	86	87	88	89	90
91	92	93	94	95	96	97	98	99	100

1. *Highlight (or color) all the even numbers in one color and all the odd numbers in another.*

2. *How many even numbers are there between 1 and 100?* ______

3. *How many odd numbers are there between 1 and 100?* ______

Roman Numerals I = 1, V = 5, X = 10, L = 50, C = 100

4. *What is the value of these Roman Numerals?*

II ______ VI ______ VIII ______ XV ______

XXX ______ XVI ______ XIX ______ LX ______

LXV ______ CXXV ______ CLXVI ______ CLXXIX ______

5. *Write the following in Roman Numerals.*

4 ______ 7 ______ 12 ______ 21 ______

17 ______ 34 ______ 39 ______ 52 ______

86 ______ 99 ______ 126______ 366 ______

6. *List these competitors in order from first to last.*

<u>Triple Jump Final</u>

Grant	9.85 m
Mark	9.21 m
John	8.90 m
Sean	9.86 m
Michael	8.78 m
Stuart	9.09 m
Robert	9.29 m

7. *Use the number grid above. Make a list of all the prime numbers between 1 and 50.*

8. *Use a* **X** *to show the square numbers to 36 on the grid.*

Assignment 12 – Money 1

1. *Change the following to $ and ¢.*

110¢ = ______ 195¢ = ______ 220¢ = ______

545¢ = ______ 1,010¢ = ______ 2,540¢ = ______

2. *What is the difference between:*

$1.00 and 25¢ = ______ $2.00 and 80¢ = ______

$2.50 and $1.20 = ______ $5.00 and $2.50 = ______

$5.00 and $3.95 = ______ $6.20 and $3.90 = ______

3. *How much change from $5.00 if I spent the following amounts:*

$1.00 = ______ $2.50 = ______ $4.95 = ______ $1.25 = ______

4. *How much change from $20.00 if I spent:*

$5.00 = ______ $2.00 = ______ $10.00 = ______ $16.00 = ______

$11.00 = ______ $19.00 = ______ $15.50 = ______ $10.80 = ______

$13.50 = ______ $8.20 = ______ $4.25 = ______ $1.90 = ______

5. *Use the takeout menu to answer the following questions.*

Lee's Chinese Takeout Menu

Hot and Sour Soup $2.50
Egg Drops Soup $2.50
Spring Rolls (4) $2.60
Fried Wontons (6) $3.00
Fried Rice (lge) $4.50
Fried Rice (sm) $3.20
Steamed Rice $2.00

Garlic Chicken $7.00
Honey Chicken $7.50
Hunan Beef $6.90
Satay Beef $7.50
Garlic Seafood $10.00
Beef Chop Suey $7.00
Veg Omelette $6.00

(a) *What three dishes could you order for under $20.00?* ______

(b) *What would it cost to order Hot and Sour Soup, Steamed Rice and Hunan Beef?* ______

(c) *What is the difference in price between a large Fried Rice and a small Fried Rice?* ______

(d) *What is the most expensive item on the menu?* ______

(e) *Four people ordered a large Fried Rice, four Spring Rolls, Honey Chicken, Satay Beef, Garlic Seafood, and Beef Chop Suey. What is the total bill?* ______

6. *Use the newspaper to draw items that you might buy if you had the following amounts to spend?*

(a) $10.00 (b) $50.00 (c) $100.00

Assignment 13 – Money 2

Carter's Family Store

For all your family's needs.

Compact discs	$25	Hand towels	75¢
Ladies shirts	$10	Video tapes	$6 each
Men's jeans	$50	Cassette tapes	$2.50 each
Novels	$12	Calculators	$9.50
Towels	$5.50	Shorts(child)	$5
Sheets(single) $15 (double) $22		Coloring books	$2.75

1. Use the above advertisement to answer the following:

(a) How much would it cost to buy four towels? ______

(b) What would be the price of two videotapes, one compact disc and three cassette tapes? ______

(c) Find the total cost of buying one of each of the clothing items. ______

(d) What is the difference in price between the single and double sheet sets? ______

(e) What is the total cost of eight coloring books and one calculator? ______

2. Make a list of items that you could buy if you had:

$75 to spend ______

______

$42.75 ______

$275.95 ______

______

$359.20 ______

______

3. Margaret had $178.20 saved for the purchase of a bike which costs $225. How much does she need to save? ______

4. One car tire costs $89. How much would it cost for four tires? ______

5. Five children each had $12.80 to spend. How much is that altogether? ______

6. A real estate agent sold one house for $89,500 and another for $110,250. What was the total price of both houses? ______

7. A deposit of $22,450 was put on a house. If the total price was $133,875, how much did the purchasers need to borrow? ______

8. Four cars were purchased for a business. If each car was $18,500, what was the total price? ______

9. Subtract $225,500 from half a million dollars. ______

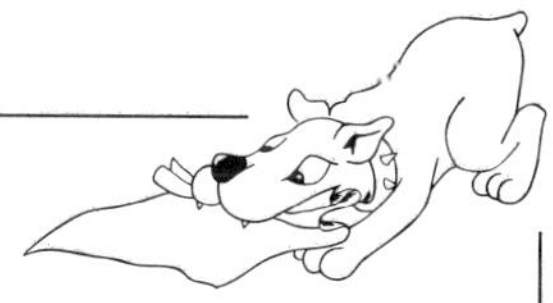

Assignment 14 – Fractions

1. *What fraction is shaded?*

(a) = ____________

(b) = ____________

(c) = ____________

(d) = ____________

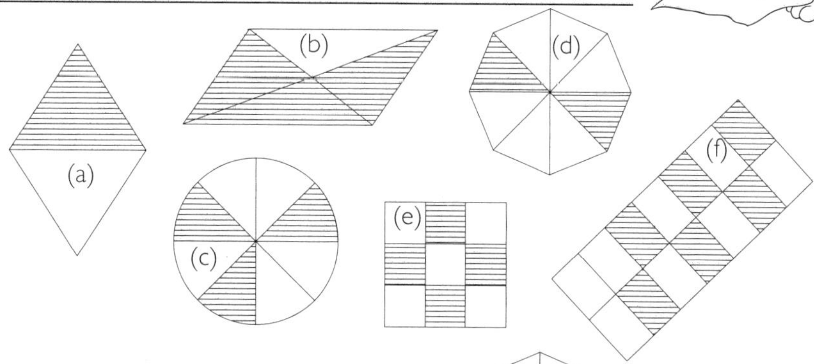

2. *Shade each to the fraction.*

(a) = 1 (d) = $\frac{5}{8}$

(b) = $\frac{1}{4}$ (e) = $\frac{8}{9}$

(c) = $\frac{7}{8}$ (f) = $\frac{9}{14}$

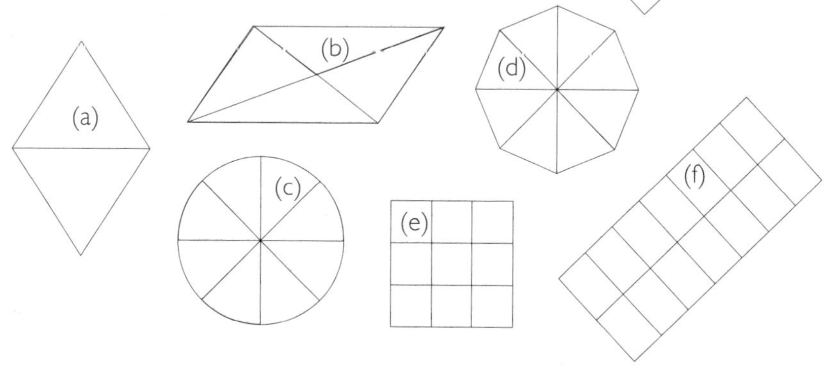

3. *Use the diagram to complete the following.*

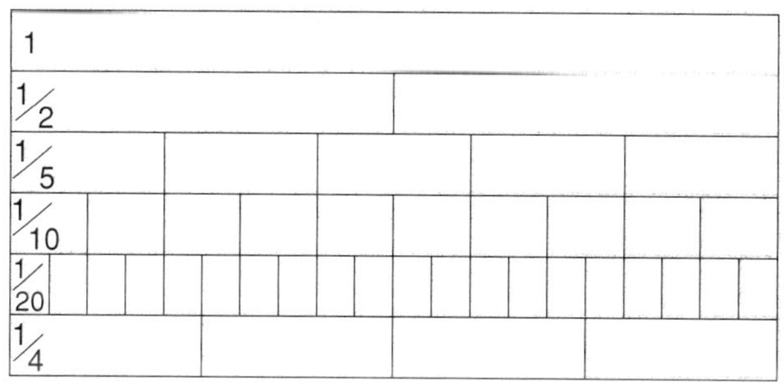

$1 = \frac{\quad}{2} = \frac{\quad}{4} = \frac{\quad}{10} = \frac{\quad}{20}$

$\frac{1}{2} = \frac{\quad}{4} = \frac{\quad}{10} = \frac{\quad}{20}$

$\frac{1}{4} = \frac{\quad}{20}$

$\frac{1}{5} = \frac{\quad}{10} = \frac{\quad}{20}$ $\frac{1}{5} = \frac{\quad}{20}$ $\frac{1}{10} = \frac{\quad}{20}$ $\frac{2}{5} = \frac{\quad}{20}$

$\frac{2}{10} = \frac{\quad}{20}$ $\frac{1}{4} = \frac{\quad}{20}$ $\frac{2}{4} = \frac{\quad}{20}$ $\frac{\quad}{5} = \frac{12}{20}$

4. *Add the following fractions.*

$\frac{1}{4} + \frac{1}{4} =$ $\frac{1}{4} + \frac{2}{4} =$ $\frac{1}{2} + \frac{1}{2} =$ $\frac{3}{5} + \frac{1}{5} =$

5. *Subtract the following fractions.*

$\frac{3}{4} - \frac{1}{4} =$ $\frac{6}{7} - \frac{2}{7} =$ $\frac{4}{5} - \frac{1}{5} =$ $\frac{6}{9} - \frac{3}{9} =$

6. *Write < or > for each.*

$\frac{1}{2}$ ______ $\frac{1}{4}$ $2\frac{7}{8}$ ______ $2\frac{2}{6}$ $1\frac{2}{3}$ ______ $1\frac{3}{4}$ $2\frac{7}{8}$ ______ $2\frac{2}{6}$

Assignment 15 – Problem Solving

Use the map to solve the following.

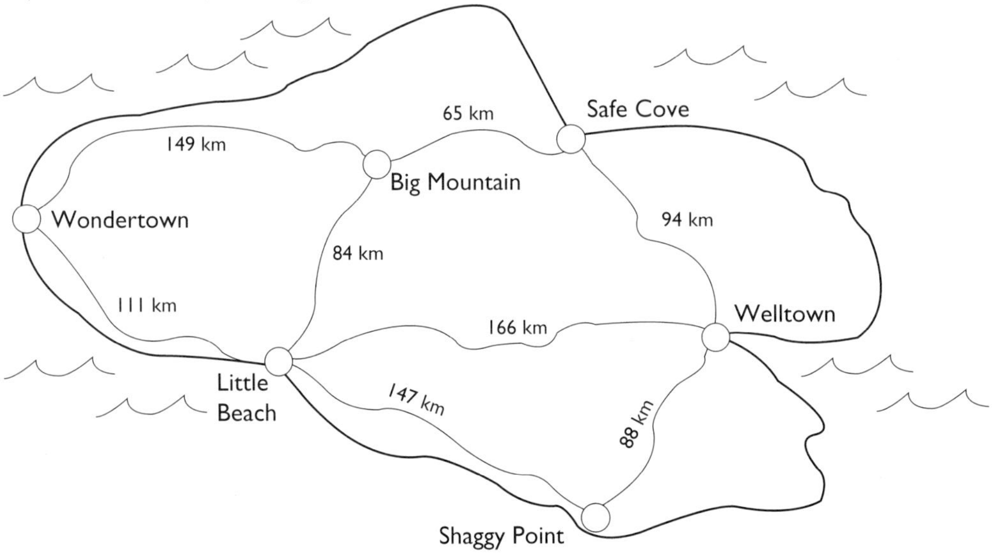

1. (a) *We traveled from Welltown to Safe Cove and back five times during our vacation. How many kilometers is this?* ____________________

(b) *How far did we travel if we went from Welltown to Shaggy Point to Little Beach and directly back to Welltown?* ____________________

(c) *Which is further from Welltown – Shaggy Point or Safe Cove?* ____________________

(d) *What is the difference between the distance from Big Mountain to Wondertown and Big Mountain to Little Beach?* ____________________

(e) *What is the total distance if we traveled from Wondertown to Big Mountain, on to Little Beach, across to Shaggy Point, up to Welltown, and on to Safe Cove?* ____________________

2. *Maria had $255.75 in her savings. She withdrew $66.80 on her birthday. How much did she have left?* ____________________

3. *A school purchased 12 footballs at $29.95 each. What was the total cost?* ____________________

4. *Elizabeth measures 154 cm, Tania is 166 cm, Nancy is 159 cm, and Peta is 162 cm. What is their combined height?* ____________________

What is the difference between the tallest and shortest? ____________________

5. *The following math scores were recorded during the semester – 75%, 82%, 100%, 78%, 92%, 95%.*

What is the total of these scores? ____________________

What is the average of these scores? ____________________

6. *Use the numbers 1, 2, 3, 4, 5, 6, 7. Place each number in a circle so each line adds up to 12.*

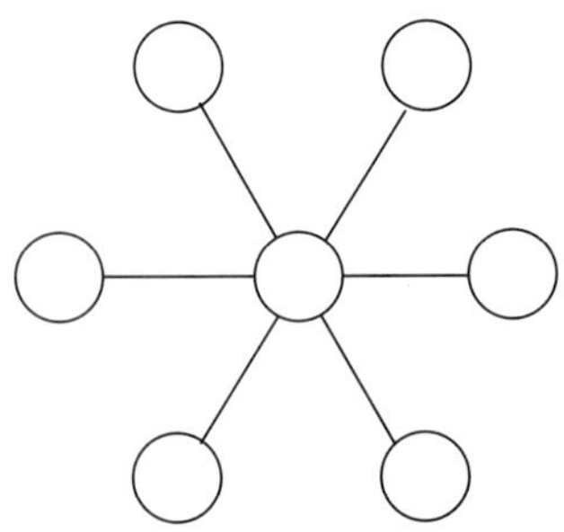

Assignment 16 – Calculators

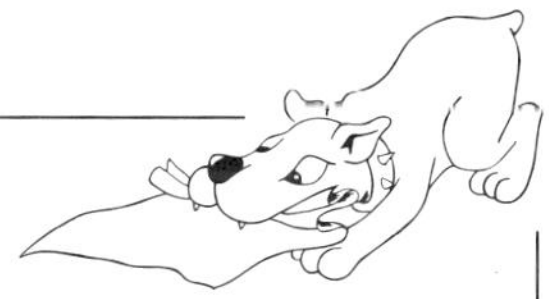

Use your calculator to solve the following problems.

1. Add the following shopping lists.

toothpaste	$2.25
toothbrush	$3.29
razor blades	$7.92
deodorant	$3.40
shampoo	$4.99
conditioner	$5.29

Total ______________

t-bone steak	$7.98
pork roast	$6.99
lamb chops	$4.99
chicken fillets	$10.29
sausages	$5.95
pork chops	$8.25

Total ______________

2 kg carrots @	$1.45 kg
5 kg potatoes @	$1.99 kg
2.5 kg oranges @	$3.90 kg
5 mangoes @	$1.29 ea
4.5 kg apples @	$2.90 kg
6 watermelons @	$2.20 ea

Total ______________

teabags	$2.20
coffee	$5.39
milk	$2.19
bread	$1.28
cheese	$4.22
soft drink (carton)	$8.99

Total ______________

2. Five towns had the following populations – 21,456; 18,909; 28,706; 22,555; 29,876. What was the total population of the five towns? ____________________

3. Which numbers in each set add to 20,000?

5,789; 2,222; 15,004; 9,850; 2,139; 2,999 = ______________________________

2,001; 5,555; 1,569; 7,124; 14,248; 5,752 = ______________________________

4. In a factory, 96 items were packed into each of 128 cartons. How many items were packed altogether? ____________

5. Find the missing number.

215 × ______ = 1,075   364 × ______ = 2,912

289 × ______ = 87,278   429 × ______ = 5,577

6. What number is multiplied by itself to give the following numbers?

__________ = 3,025   __________ = 7,569   __________ = 8,281

__________ = 12,100   __________ = 2,401   __________ = 5,929

__________ = 251,001   __________ = 20.25   __________ = 650.25

7. The library had 50,000 books. If they gave away 2,987, how many did they have left? ____________

8. Add the following amounts – one million, half a million, 42,569, one-quarter of a million, 7.98, 500.50, and 6. ____________________

Assignment 17 – Codes

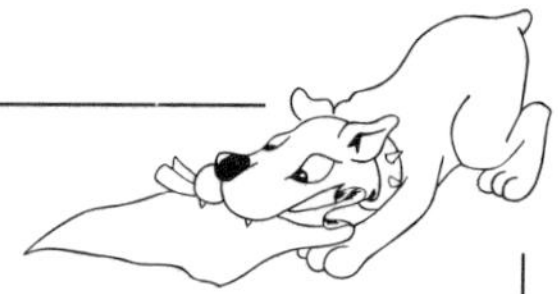

1. *Use the code to read the message.*

A	B	C	D	E	F	G	H	I	J	K	L	M	N	O	P	Q	R	S	T	U	V	W	X	Y	Z
1	2	3	4	5	6	7	8	9	10	11	12	13	14	15	16	17	18	19	20	21	22	23	24	25	26

1 13 5 18 9 3 1   9 19   1   7 18 5 1 20   16 12 1 3 5
20 15   12 9 22 5.

____

____

Answer the following question using the code.

23 8 9 3 8   19 20 1 20 5   4 15   25 15 21   12 9 22 5   9 14?

____

2. *Complete the following magic squares.*

	7	
	9	
10	11	6

6		4
10	3	8

12		
	15	
24	3	18

9	8	7
5		

		18
	20	
22		17

3. *Complete these problems by finding the missing numbers.*

```
  5 _ 9        6,2 1 _         9 8 2        5 6 3
+ 2 4 _     + _,4 9 6          _ _ 4        2 1 _
-------     ---------        + 6 4 _      + _ 6 _
  _ 8 1      9,_ _ 0         -------      -------
                              2,1 8 9      1,7 4 3

  8 6 _      6,_ 8 7            5 6          6 _ 4
+ _ 5 7    - _,8 2 _          ×   _        ×     _
-------    ---------          -----        -------
  4 _ 1      3,1 _ 4          1 _ 8        4,_ 3 8
```

4. *Answer these questions using the code from question one.*

What is your name? ______________________________

What school do you attend? ______________________________

Assignment 18 – Measuring

1. *Complete these.*

1 cm = ____ mm 1 m = ____ cm 1 km = _____ m

2. *Draw an item in each box that would be measured with that unit of measurement.*

METERS	CENTIMETERS	MILLIMETERS

3. *Draw lines that measure about* 2 cm, 5 cm, 10 cm, 15 cm.

Use your ruler to check the actual measurements.

4. *Change the following to centimeters.*

3 m = ____________ 2 m = ____________ 4.5 m = ____________

10 m = ____________ 2.6 m = ____________ 3.5 m = ____________

4.85 m = ____________ 8.6 m = ____________ 6.95 m = ____________

2.08 m = ____________ 7.7 m = ____________ 9.17 m = ____________

5. *Change the following to meters.*

200 cm = ____________ 800 cm = ____________ 1,000 cm = ____________

358 cm = ____________ 49 cm = ____________ 440 cm = ____________

6. *Find the difference in length between the following.*

1 m and 50 cm = ______________ 2 m and 400 cm = ______________

5.40 m and 40 cm = ______________ 425 cm and 20 cm = ______________

10 m and 990 cm = ______________ 11.8 m and 1,100 cm = ______________

7. *Cathy is* 1.55 m *tall and Tanya is* 1.67 m *tall. How much taller is Tanya?* ______________

8. *Four boys recorded jumps of* 2.94 m, 2.88 m, 3.09 m, *and* 3.32 m *in the long jump. What was the total distance jumped?*(a) ______________

What is the difference between the shortest and longest jumps? (b) ______________

If the fifth boy won the event with a jump of 3.56 m, *what was the difference between first and second place?* (c) ______________

Assignment 19 – Perimeter

1. *Use a ruler to measure the perimeters of these shapes.*

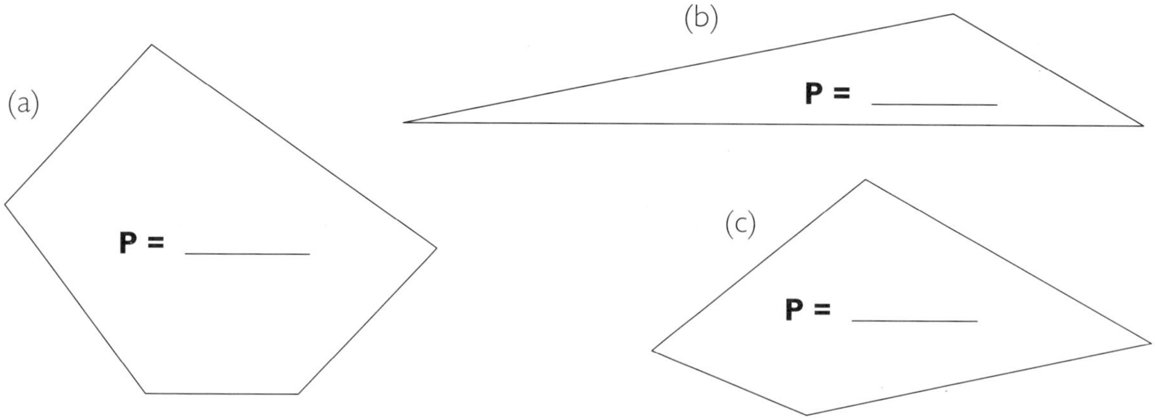

2. *Complete the table.*

length	width	perimeter
4 cm	2 cm	**12 cm**
10 mm	5 mm	**30 mm**
25 m	10 m	**70 m**
5.5 m	3 m	**17 m**
12.5 cm	6.5 cm	**38 cm**

3. *Draw a shape with five sides. Measure each side in mm and then find the perimeter.*

4. *On the grid draw as many shapes as you can with a perimeter of 24.*

5. *Find the perimeter of a book if the length is 22 cm and the width is 18 cm.* ____________

6. *Find the perimeter of a garden bed if the length is 4.2 m and the width is 1.7 m.* ____________

7. *Find the perimeter of your notebook.* ____________ *Your desktop.* ____________

Assignment 20 – Area

1. *Count the squares to find the area of the following shapes.*

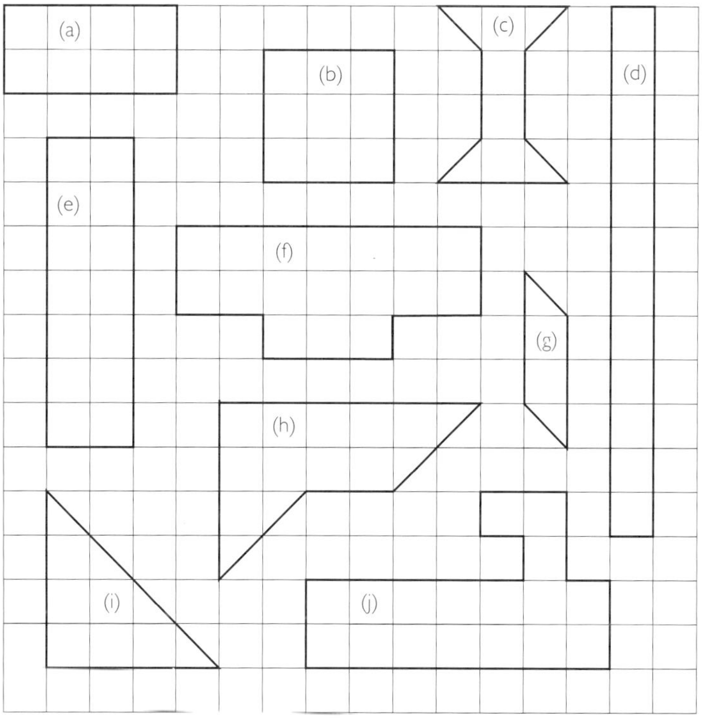

(a) A = ________

(b) A = ________

(c) A = ________

(d) A = ________

(e) A = ________

(f) A = ________

(g) A = ________

(h) A = ________

(i) A = ________

(j) A = ________

2. *Draw the following shapes on graph paper and find the area of each.*

(a) square – 4 cm long A = ________ (c) rectangle – 6 cm long, 3 cm wide A = ________

(b) square – 6 cm long A = ________ (d) rectangle – 8 cm long, 4.5 cm wide A = ________

3. *On the grid below, draw five different shapes that have an area of 8 square units.*

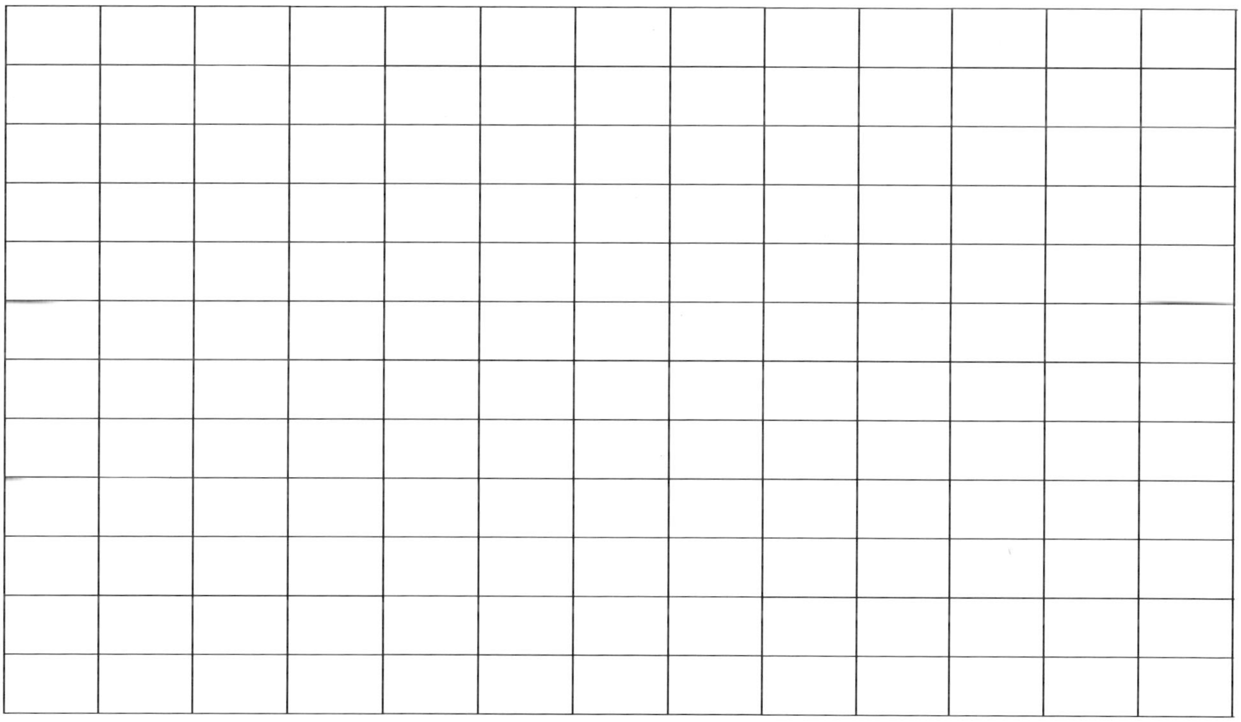

4. *Draw a line to cover 36 squares (any shape).*
Color a pattern of 8 squares in blue, 12 squares in green and 16 squares in red.

Assignment 21 – Mass

1. *How many kilograms in one kilogram?* ____________

2. *How many kilograms in one ton?* ____________

3. *Complete the table by listing the following items in the column that shows the best unit of measurement.*
truck, computer, pen, two marbles, adult, three bricks, small doll, library book, four gold bars, empty lunch box, toothpick

grams	kilograms	tons

4. *Change the following to kilograms.*

1,000 g = ____________ 3,000 g = ____________ 2,250 g = ____________
4,000 g = ____________ 8,426 g = ____________ 3,415 g = ____________
2,005 g = ____________ 2,098 g = ____________ 3,456 g = ____________

5. *Change the following to grams.*

2 kg = ____________ 4 kg = ____________ 3.5 kg = ____________
3.8 kg = ____________ 5.25 kg = ____________ 10 kg = ____________
12.4 kg = ____________ 16.8 kg = ____________ 13.55 kg = ____________
5.8 kg = ____________ 11 kg = ____________ 6.8 kg = ____________

6. *How much do you weigh?* __________ kg

7. *If you and four of your friends weighed the same, what would be the total mass of the five?* ____ x ____ = ____

8. *Find the difference between:*

1 kg and 500 g = ____________ 1,000 g and 250 g = ____________
2.5 kg and 1 kg = ____________ 2,500 g and 2 kg = ____________
6,000 g and 5 kg = ____________ 6,027 g and 3.006 kg = ____________

9. *Find the mass of the following people – Sandy – 42 kg, Shelly – 44 kg, Robert – 45 kg, and Pam – 45 kg.* ____________

10. *Find the mass of the people who live in your house. Show your information as a bar graph.*

Assignment 22 – Tallies

1. *Make a tally to show how many of the letters A – K are found below.*

B	A	J	A	A	F	E	B	E	J	H	H	I	I	A	B	C	D	E
H	I	J	K	J	K	B	A	D	E	F	A	A	I	J	J	J	K	A
B	A	G	G	A	C	G	H	J	K	K	K	E	A	B	D	E	G	F
H	I	K	A	B	D	F	E	G	I	H	A	C	A	J	K	A	A	I

(a) *Which letter occurs most frequently?* ______________

(b) *Which letter occurs least frequently?* ______________

(c) *What is the total number of letters?* ______________

2. *Make a tally to show the type of advertisements on TV between* 4 p.m. *and* 5 p.m.

Advertisement	Tally	Total
food		
toys		
cleaning products		
cars		
clothing		

(a) *Which advertisement is the most popular?* ______________________________

(b) *Which advertisement is the least popular?* ______________________________

3. *Make a tally to show the types of utensils in your drawer at home.*

Item	Tally	Total

(a) *What is the total number of utensils in your drawer?* ________________

(b) *What do you have the most number of?* ____________________

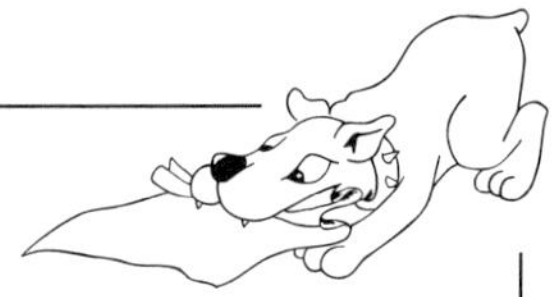

Assignment 23 – Graphs

1. *Use the bar graph to answer the following questions.*

(a) *How many children are in Room 12?* ______________________

(b) *Which was the least popular drink?* ______________________

(c) *Which was the most popular drink?* ______________________

(d) *How many children preferred orange?* ______________________

(e) *How many more children preferred cola to raspberry?* ______________________

(f) *How many children preferred another drink to lemonade?* ______________________

(g) *Which drinks received the same number of votes?* ______________________

(h) *How many children chose orange, lemon and ginger ale?* ______________________

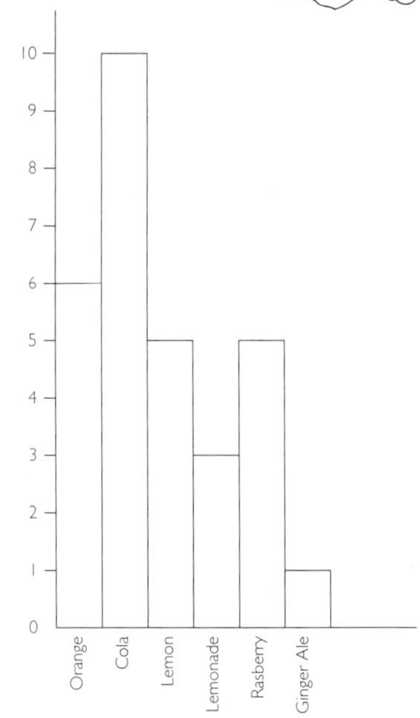

2. *Use the following information to construct your own bar graph on the grid provided.*

EYE COLOR

Color	No. of children
Blue	12
Green	7
Brown	10
Hazel	2

3. *Use the following information to construct a line graph.*

Maximum Temperatures

Week 1		Week 2	
Mon	28	Mon	30
Tue	29	Tue	32
Wed	25	Wed	28
Thur	27	Thur	34
Fri	28	Fri	33

Assignment 24 – Time 1

1. *Draw hands on the clock faces to show the times underneath.*

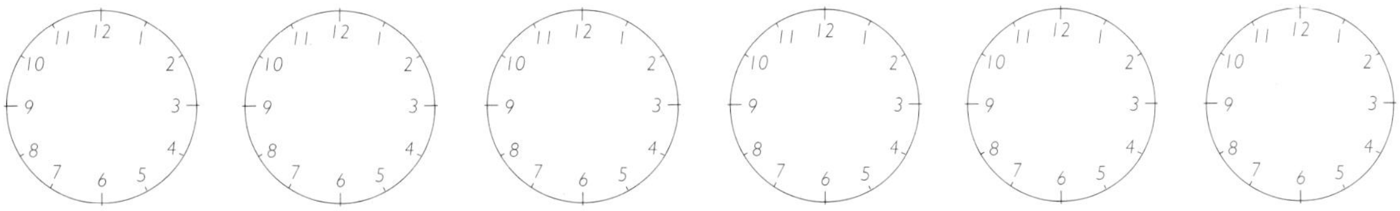

7:30 12:00 2:15 8:45 9:20 10:10

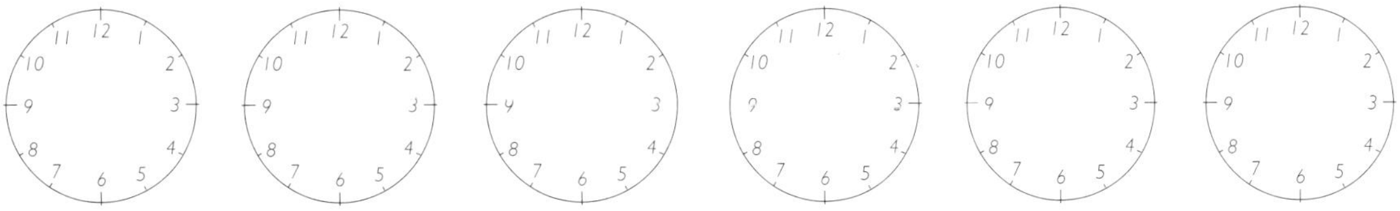

11:25 12:45 3:40 1:55 4:35 5:05

2. *Change these 24-hour times to 12-hour times.*

1300 ________	1530 ________	2015 ________	1445 ________
1610 ________	1920 ________	2240 ________	1800 ________
0610 ________	0101 ________	2400 ________	1200 ________

3. *Complete the following.*

______ h = 1 day ______ min = 1 h ______ min = $^1/_2$ h ______ s = 1 min

4. *How many:* *hours in 2 days?* ________ *hours in a day and a half?* ________

seconds in 4 minutes? ________ *minutes in 5 hours?* ________

minutes in 2 $^1/_2$ hours? ________ *days in 2 weeks?* ________

5. *Write down the exact time now. The time is* ________

What will be the time in: 10 minutes? ________ *in 14 minutes?* ________

What was the time: *10 minutes ago?* ________ *one hour ago?* ________

6. *Write down the times that you do the following. Use 12-hour times and 24 hour times. During a school week, what time do you:*

(a) *Wake up?*	________	________
(b) *Have breakfast?*	________	________
(c) *Have lunch?*	________	________
(d) *Go to school?*	________	________
(e) *Arrive at school?*	________	________
(f) *Go to morning recess?*	________	________
(g) *Start math on a Monday?*	________	________
(h) *Leave school?*	________	________
(i) *Arrive home?*	________	________

Assignment 25 – Time 2

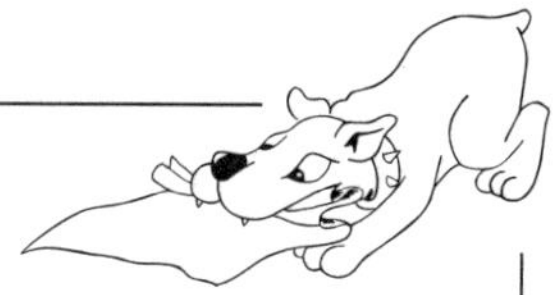

1. *Show the time that has passed between these departure and arrival times.*

(a) Depart 7:00 a.m. Arrive 9:30 a.m. ______________
(b) Depart 10:00 a.m. Arrive 1:15 p.m. ______________
(c) Depart 11:30 a.m. Arrive 3:45 p.m. ______________
(d) Depart 5:15 p.m. Arrive 10:10 p.m. ______________
(e) Depart 11:55 p.m. Arrive 8:00 a.m. ______________

(f) Depart 0800 Arrive 1020 ______________
(g) Depart 0925 Arrive 1145 ______________
(h) Depart 1200 Arrive 1535 ______________
(i) Depart 1440 Arrive 2010 ______________
(j) Depart 2400 Arrive 1350 ______________

2. *Complete your class timetable.*

TIME	MON	TUE	WED	THU	FRI
	RECESS				
	LUNCH				
	RECESS				

(a) How long do you spend at lunch each week? ______________

(b) How long do you spend in school each day (exclude lunch and recess)? ______________

(c) How long do you spend in school each day (include lunch and recess)? ______________

(d) How many hours a week do you study mathematics? ______________

(e) How many hours a week do you work on your favorite subject? ______________

(f) What times do you have physical education? ______________

3. *Brett started basketball practice at 1545 and finished at 1700. If he went three times a week, how long did he spend at practice?* ______________

4. *David left home at 0815 and arrived 3 hours and 40 minutes later. What time did he arrive?* ______________

5. *Kelly left home at 4:35 p.m. and arrived at 7:15 p.m. How long did she take to arrive at her destination?* ______________

Assignment 26 – Calendars

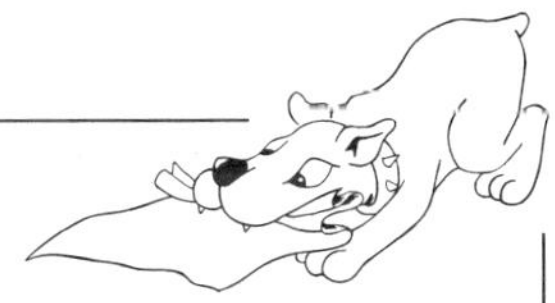

1. *How many months are there in a year?* ______________________

2. *How many days are there in a year?* __________ *leap year?* __________

3. *How many days are there in January?* __________ *February?* __________

May? __________ *September?* __________ *December?* __________

4. *On what day does Christmas fall this year?* ______________________

5. *What will be the date two weeks after your birthday?* ______________________

6. *What will be the date eight days after Christmas Day?* ______________________

7. *What will be the date three weeks after today?* ______________________

8. *How many months have 30 days?* ______________________

9. *What will be the date nine days after June 28?* ______________________

10. *Use these calendars to help you answer the following questions.*

NOVEMBER

SUN	MON	TUE	WED	THUR	FRI	SAT
						1
2	3	4	5	6	7	8
9	10	11	12	13	14	15
16	17	18	19	20	21	22
23	24	25	26	27	28	29
30	31					

DECEMBER

SUN	MON	TUE	WED	THUR	FRI	SAT
						1
2	3	4	5	6	7	8
9	10	11	12	13	14	15
16	17	18	19	20	21	22
23	24	25	26	27	28	29
30	31					

(a) *On what day is November 30?* __________

(b) *What will be the date three weeks before December 3?* __________

(c) *How many days are there between November 13 and December 3?* __________

(d) *What is the date two weeks from December 12?* __________

(e) *How many days are there between November 1 and December 2?* __________

(f) *Pay day is every second Thursday. If the first pay day is November 1, how many will there be before Christmas Day?* __________

(g) *What will be the date 13 days after November 24?* __________

11. *Label the boxes below with each of the four seasons. Write the correct months in each box.*

Assignment 27 – Shapes

1. *Choose the correct name for each shape.*

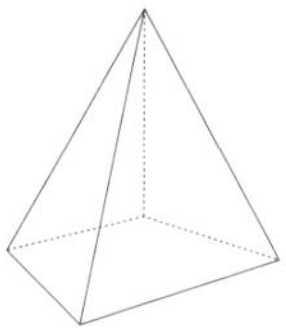

(a) ____________________

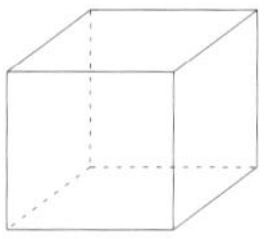

(b) ____________________

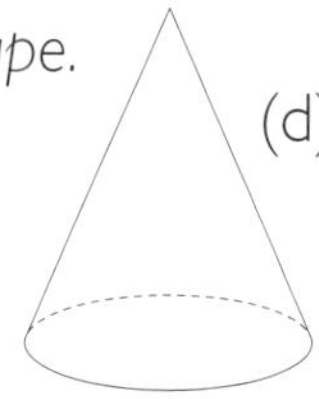

(d) ____________________

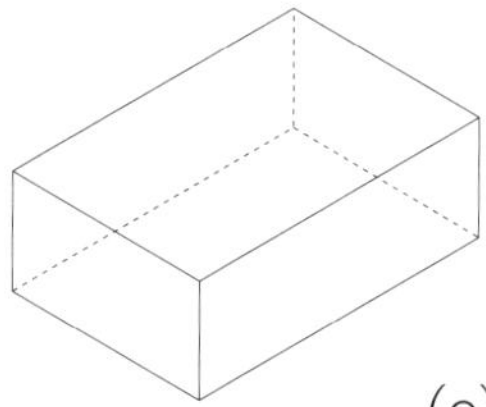

(c) ____________________

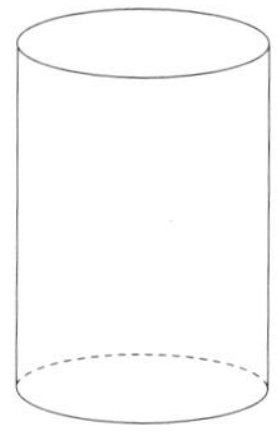

(e) ____________________

cube, cone, cylinder, square pyramid, rectangular prism

2. *Use the above shapes to complete the table.*

SHAPE	VERTICES	FACES	EDGES
cube			
cone			
cylinder			
square pyramid			
rectangular prism			

3. *Give two examples of objects that have the same shape as each of those above.*

cube – ____________________

cone – ____________________

cylinder – ____________________

square pyramid – ____________________

rectangular prism – ____________________

4. *What shape does this net represent?*

____

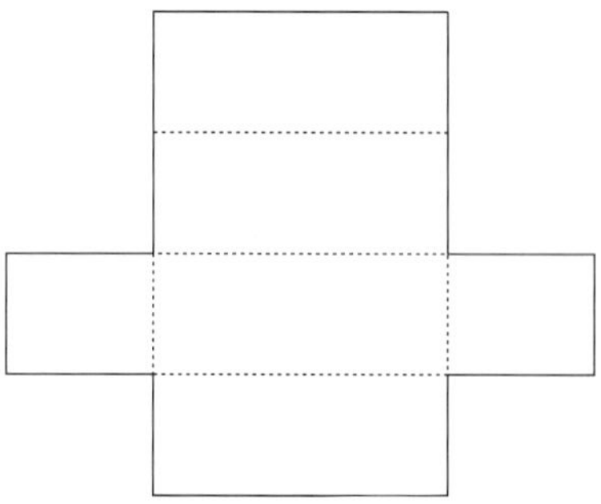

5. *Draw the net for:*

a triangular pyramid

6. *On the back of this sheet, draw two shapes you could make from straws that would be rigid.*

7. *On the back of this sheet, draw two shapes you could make from straws that would not be rigid.*

Assignment 28 – Tessellations

Example.

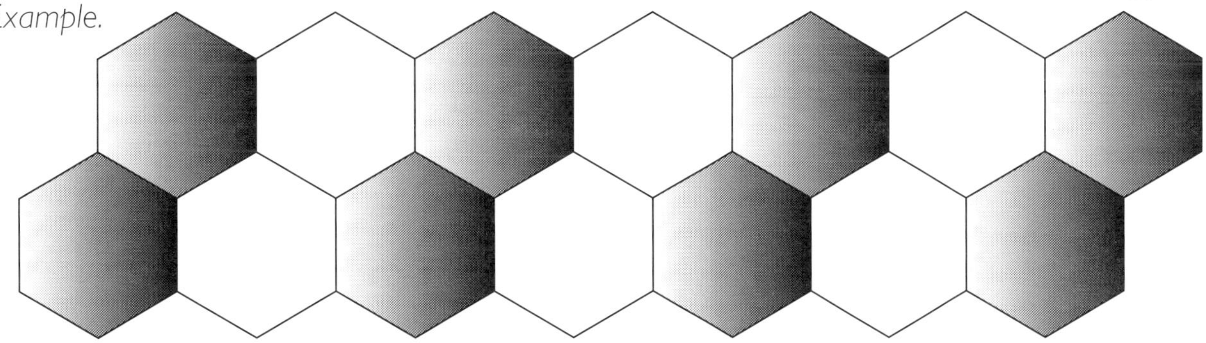

Tessellations are repeated patterns of shapes which completely cover a surface without overlapping.

In each of the six boxes choose a shape that tessellates and makes a pattern.
Color each pattern differently.

1.

4.

2.

5.

3.

6.

Assignment 29 – Reducing and Enlarging

1. *Enlarge the shapes on the grid below.*

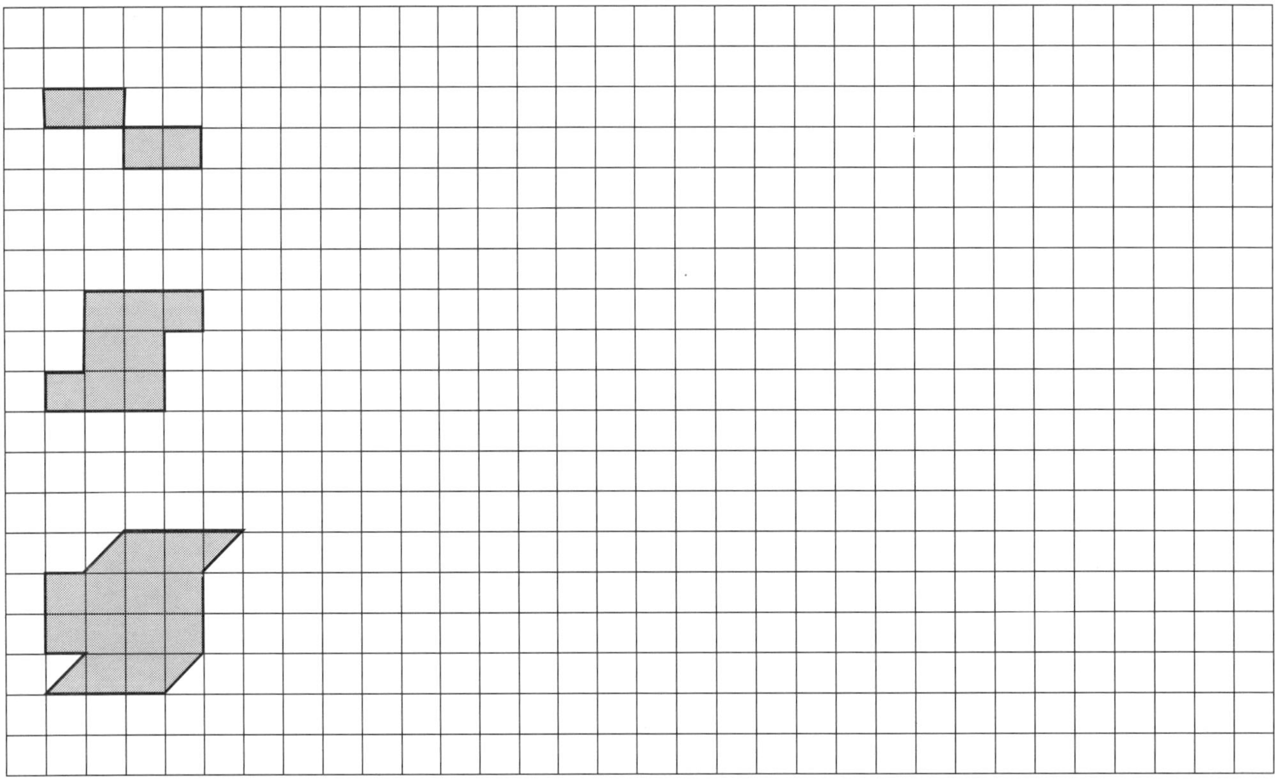

2. *Reduce the shape on the grid below.*

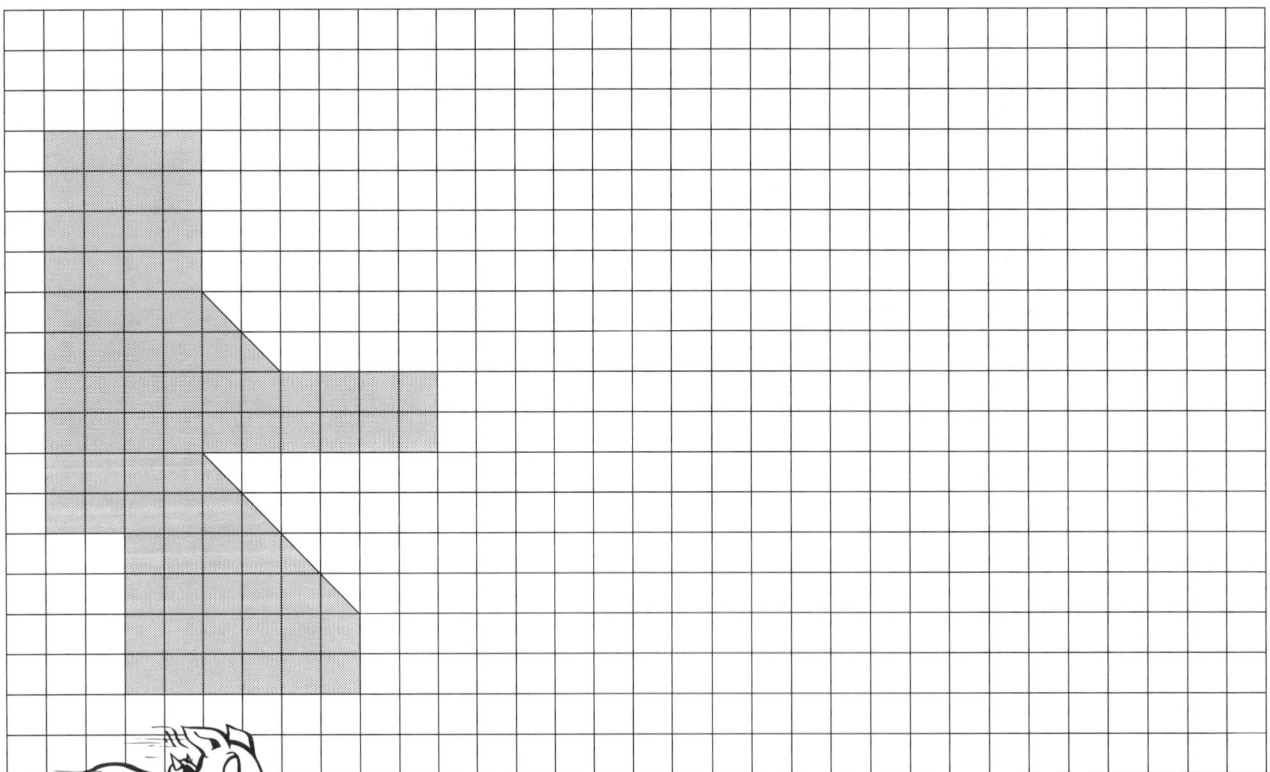

Assignment 30 – Angles

1. *Use your protractor to estimate and measure these angles.*

Estimate				Measure
Angle	**Right Angle**	**Less than Right Angle**	**More than Right Angle**	**Size**
A B C				
D E F				
G H I				
J K L				
M N O				
P Q R				

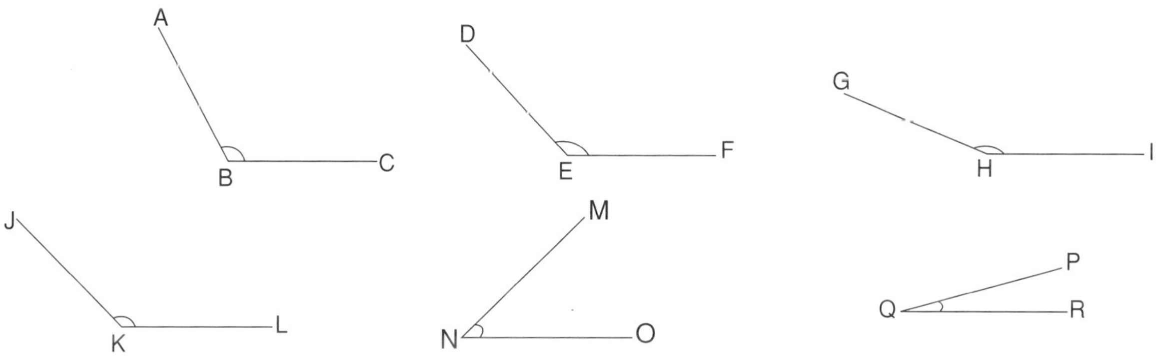

2. *Draw and label a right angle, an angle less than 90° and an angle more than 90°.*

3. *Measure the following angles.*

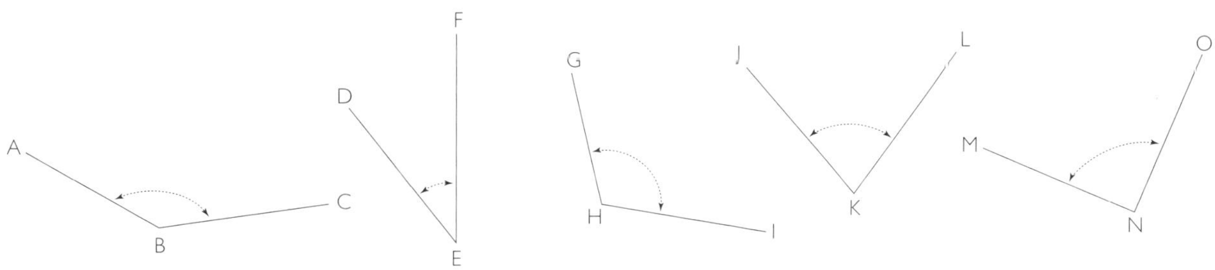

∠ABC = ______ ∠DEF = ______ ∠GHI = ______ ∠JKL = ______ ∠MNO = ______

Assignment 31 – Networks

1. *Decide which of these networks you can trace over without lifting your pencil and without retracing any line. Circle YES or NO.*

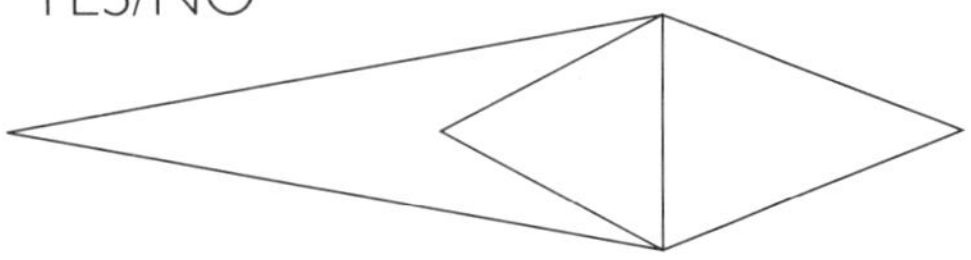

(a) YES/NO

(b) YES/NO

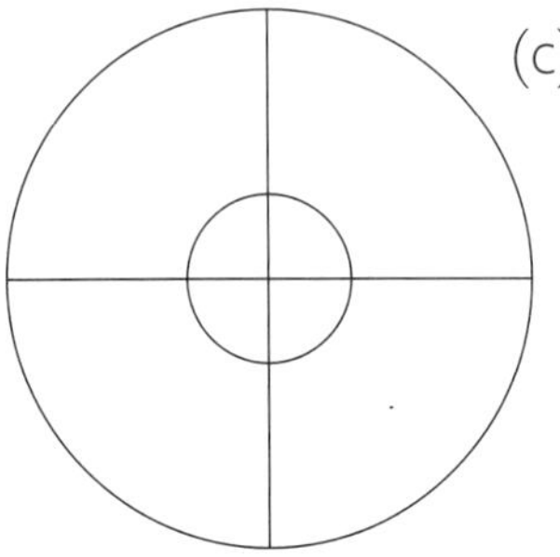

(c) YES/NO

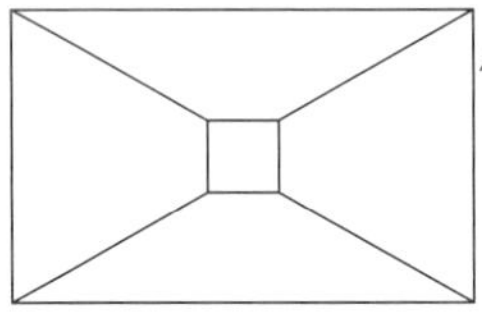

(d) YES/NO

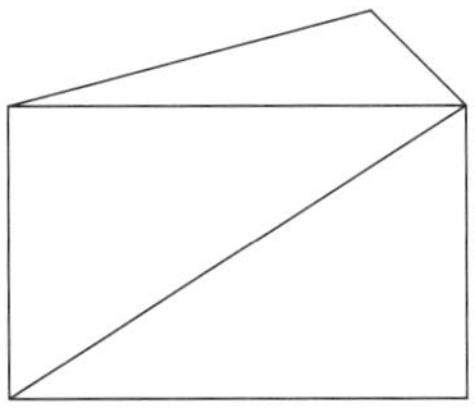

(e) YES/NO

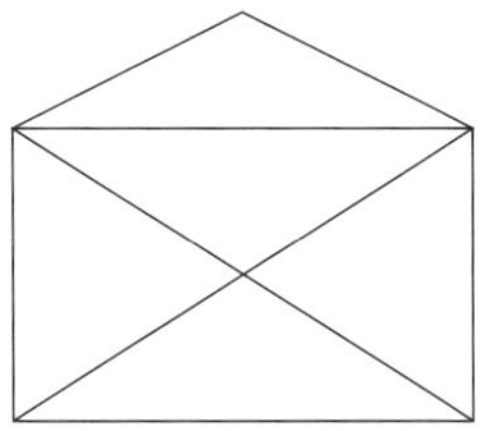

(f) YES/NO

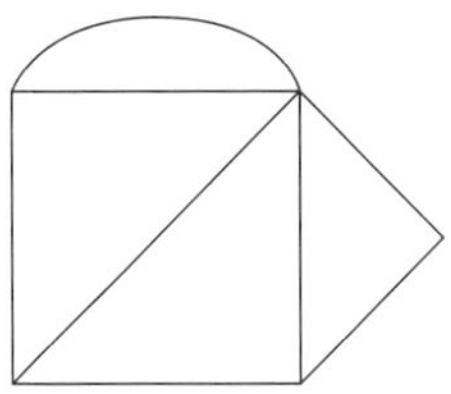

(g) YES/NO

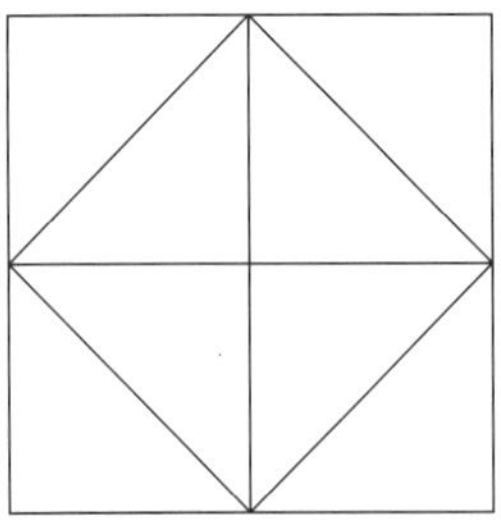

(h) YES/NO

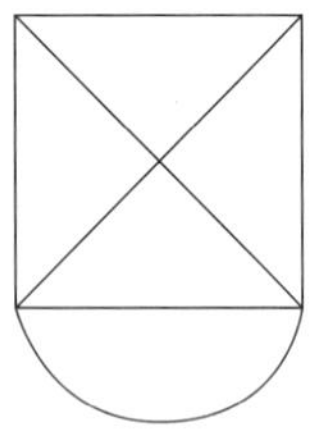

(i) YES/NO

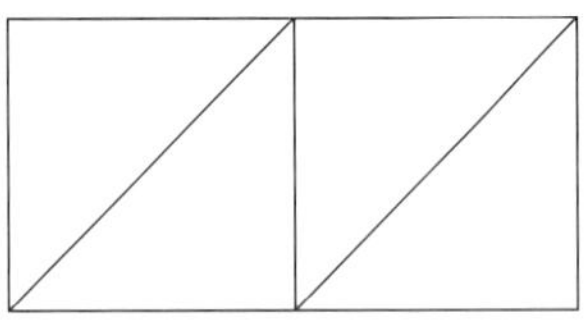

(j) YES/NO

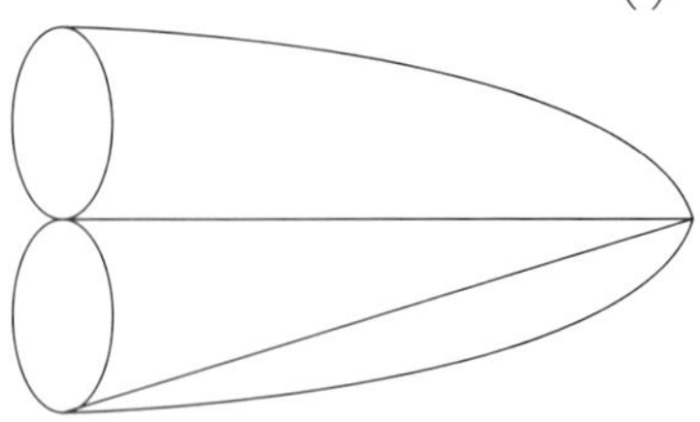

(k) YES/NO

2. *Draw some networks of your own on the back of this page. Swap with a friend and see if they can trace over your networks without lifting their pencil or retracing any lines.*

Assignment 32 – Word Search

Complete the word search by finding the words across, down, forward, backward, or diagonally. When you have found all the words the remaining letters will spell a mystery word.

S	H	A	P	E	S	E	M	A	S	E	L	G	N	A
E	F	R	A	C	T	I	O	N	S	S	S	A	M	C
L	E	L	Z	Z	U	P	N	M	E	E	S	E	E	A
B	R	E	D	U	C	E	E	U	N	T	E	N	A	L
A	A	S	E	D	O	C	Y	L	T	S	L	L	S	E
T	R	P	E	R	I	M	E	T	E	R	G	A	U	N
N	E	T	W	O	R	K	S	I	N	I	N	R	R	D
I	A	S	O	L	V	E	D	P	C	G	A	G	I	A
N	P	Y	S	S	E	N	I	L	E	I	I	E	N	R
E	C	U	B	E	V	F	N	I	S	D	R	R	G	S
T	S	I	Z	E	L	A	O	C	L	A	T	N	E	M
Y	N	E	T	C	R	A	I	A	D	E	E	M	I	T
N	M	I	I	A	N	E	T	T	N	C	N	X	M	N
O	E	E	U	L	A	V	C	I	U	I	E	N	E	O
I	R	I	A	P	L	R	A	O	O	D	V	U	L	I
T	R	A	C	E	S	A	R	N	R	N	E	M	B	S
I	T	H	G	I	A	R	T	S	U	M	S	B	O	I
D	I	V	I	D	E	S	B	U	I	L	D	E	R	V
D	S	R	O	T	A	L	U	C	L	A	C	R	P	I
A	G	R	A	P	H	S	S	L	A	M	I	C	E	D

addition	angles	arc	area	build	calculators	calendars
codes	cube	decimals	dice	divides	division	enlarge
even	fractions	graphs	lines	mass	measuring	mental
mixed	money	value	net	networks	ninety	number
pair	size	place	rigid	puzzle	reduce	problem
round	same	sentences	sets	seven	shapes	perimeter
solved	straight	subtraction	sums	tables	tallies	ten
time	traces	triangles	multiplication			

Mystery Word – ______________________________

Assignment 33 – Review I

1. *Draw the net for a square pyramid on the back of this sheet.*

2. *On the back of this sheet, draw a shape you could make from straws that would be rigid.*

3.	**4.**	**5.**	**6.**
456	1,248	3,555	20.456
+ 398	+ 365	+ 4,999	+ 17.987
______	______	______	______

7. *On the first day of a show 8,562 people attended, on the second day there were 5,405, on the third day there were 6,222 people, and on the fourth day there were 9,045. How many attended altogether?* ______________

8. *Use a protractor to measure the following angles.*

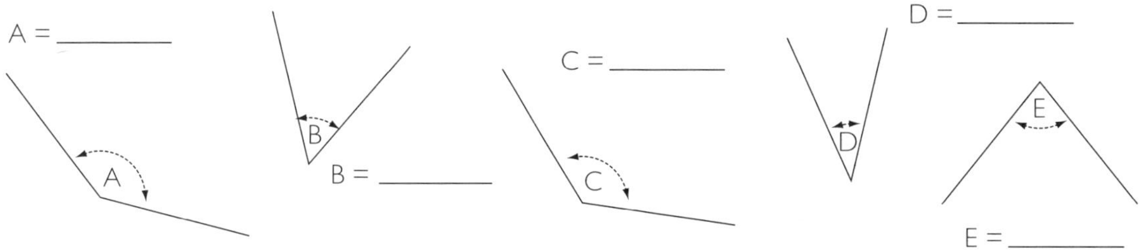

A = ________ B = ________ C = ________ D = ________ E = ________

9. *Measure all the angles in the shapes below.*

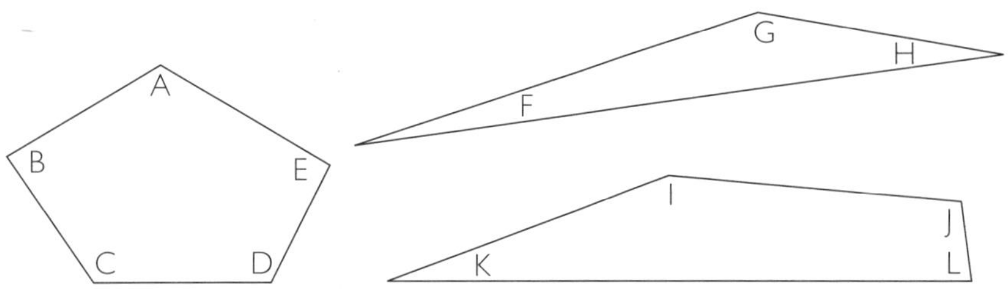

∠A = ______, ∠B = ______,
∠C = ______, ∠D = ______,
∠E = ______ ∠F = ______,
∠G = ______, ∠H = ______ ∠I = ______, ∠J = ______, ∠K = ______, ∠L = ______

10. *Find the area of a square if one side measures 5 cm.* ______________

11. *Find the area of a rectangle if the length is 10 cm and the width is 8 cm.* ______________

12. *Put these decimals in order from smallest to largest.*

3.5, 3.0, 3.9, 3.1, 3.7, 3.8, 3.6 ______________________________

15.6, 15.3, 15.0, 15.1, 15.8, 15.4 ______________________________

13. *What will be the date four weeks after today?* ______________

14. *What will be the date three weeks and two days before Christmas Day?* ______________

15. *What will be the date five weeks and one day after January 1?* ______________

16. *Complete the tables.*

4 × 8 = ______	7 × 7 = ______	12 × 5 = ______	9 × 6 = ______
10 × 7 = ______	5 × 5 = ______	8 × 4 = ______	6 × 6 = ______
11 × 8 = ______	9 × 9 = ______	6 × 8 = ______	4 × 9 = ______

Assignment 34 – Review 2

1. *Divide 40 cards among 5 people.* ______________

2. *Share $88 among 8 people.* ______________

3. *Eight children spent $52 on movie tickets. How much was each ticket?* ______________

4. *Four buses took 216 children to the swimming pool. How many children were on each bus?* ______________

5. *Add the following fractions.*

$\frac{2}{5} + \frac{1}{5} =$ $\frac{4}{20} + \frac{11}{20} =$ $\frac{6}{8} + \frac{1}{8} =$

6. *Shade each to show the fraction.*

$\frac{1}{2}$

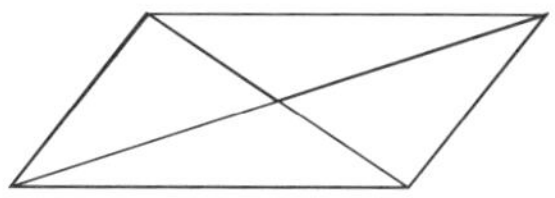

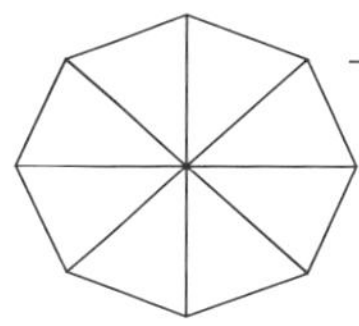

$\frac{2}{8}$

$\frac{5}{8}$

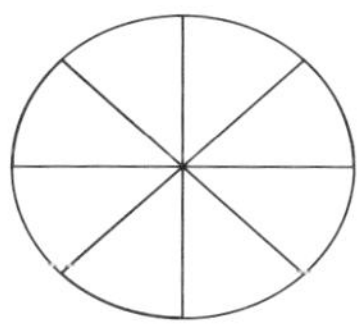

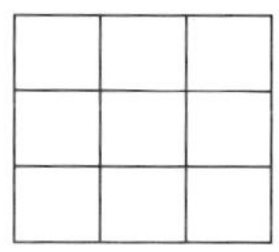

$\frac{6}{9}$

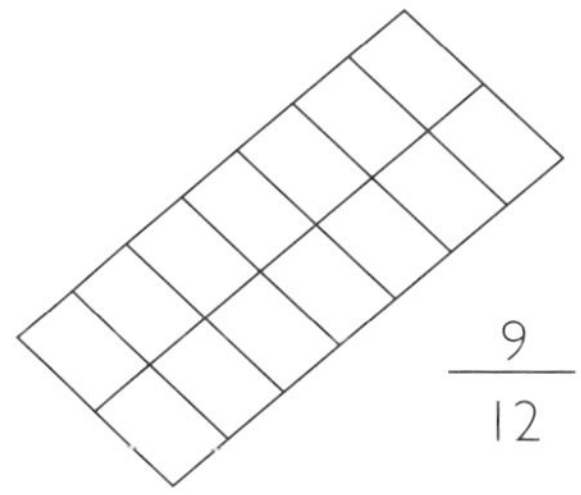

$\frac{9}{12}$

7. *Complete the following.*

$8 + 3 =$ ______ $7 \times 3 =$ ______ $9 \div 9 =$ ______ $10 - 2 =$ ______

$9 \times 5 =$ ______ $8 \times 8 =$ ______ $12 \div 4 =$ ______ $19 + 9 =$ ______

8. *Write the following numbers using digits.*

sixty-eight = ______________ *one hundred and fifty-six* = ______________

two thousand nine hundred and nine = ______________

twenty thousand and three = ______________

one hundred and seventy-three thousand four hundred and twelve = ______________

one million = ____________________________

9. *Write the place value of the 2 in each of these.*

82 = ______ 23 = ______ 209 = __________ 2,345 = ______

21,345 = ______ 210,345 = __________ 2,000,000 = ______________

10. 1 kg = ___________ g 1 ton = ___________ kg

11. *Change to kilograms.*

1,000 g = _____ kg 5,000 g = _____ kg 2,500 g = _____ kg 10,500 g = _____ kg

12. *Change to grams.*

3 kg = ________ g 4.5 kg = ________ g 7.9 kg = ________ g 10.25 kg = ________ g

13. *Change to centimeters.*

4 m = _____ cm 3.5 m = _____ cm 10 m = ________ cm 5.55 m = _____ cm

14. *Change to meters.*

300 cm = _____ m 600 cm = _____ m 550 cm = _____ m 985 cm = _____ m

Assignment 35 – Review 3

1.	**2.**	**3.**	**4.**	**5.**	**6.**
42 × 2	52 × 5	36 × 5	134 × 6	66 × 50	300 × 38

7. *If one carton holds 12 cans, how many cans will be in 15 cartons?* ________

8. *One bus pass was \$10.50. How much are four passes?* ________

9. *Draw a network that can be traced over without lifting your pencil or retracing any line.*

10. *Write true or false for each of these.*

$7 \times 8 = 60 - 4$ ________ $12 \times 3 < 30 + 5 + 2$ ________

$(5 + 3) \times 3 > 7 \times 4$ ________ $50\% > \frac{1}{2}$ ________

$10 + 10 + 10 < 10 \times 3 + 10$ ________ $75\% = \frac{3}{4}$ ________

11. *Complete these number sentences.*

1, 3, 5, ___, ___, ___, ___ 10, ___, ___, ___, 50, ___

7, 14, ___, ___, ___, ___, 49 2, 4, 3, 5, 4, ___, ___, ___, ___, ___

12. *What is the difference between the following?*

\$2.00 and 75¢ ________ \$3.00 and \$1.50 ________

\$5.00 and \$3.20 ________ \$10.00 and \$5.20 ________

\$10.35 and \$3.95 ________ \$8.25 and \$4.75 ________

13. *How much change from \$50.00 if I spent the following amounts?*

\$5.00 ______	\$15.00 ______	\$25.00 ______	\$31.00 ______
\$49.00 ______	\$20.50 ______	\$31.75 ______	\$42.80 ______
\$14.45 ______	\$6.25 ______	\$26.30 ______	\$34.15 ______

14. *One airfare was \$1,295. How much for two fares?* ________

15. *Add \$2,456.95 and \$7,899.59.* ________

16. *Measure the perimeters of these shapes.*

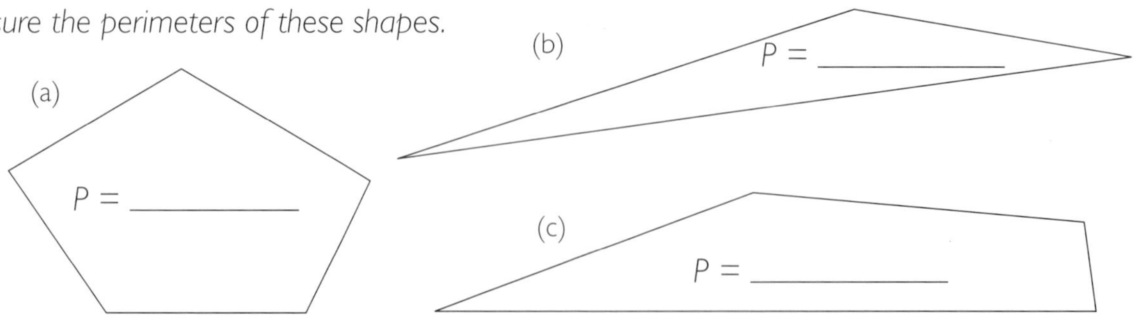

17. *Find the perimeter of a rectangle if the length is 18.4 cm and the width is 9.54 cm.*
P = ________

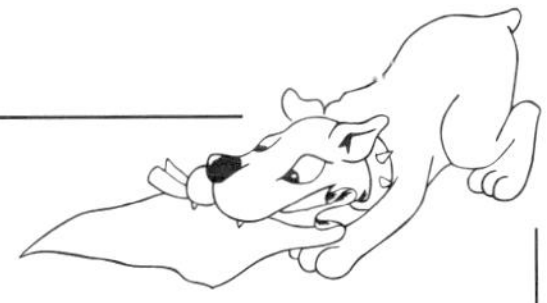

Assignment 36 – Review 4

1. *Round these numbers to the nearest 10.*

42 ______ 89 ______ 124 ______ 298 ______ 356 ______

899 ______ 1,209 ______ 2,345 ______ 2,556 ______ 409 ______

2. *Round each number to the nearest 10 before adding.*

42 + 78 + 56 + 91 + 99 ~ ______________________ = ______

9 + 78 + 98 + 104 + 67 ~ ______________________ = ______

38 + 29 + 71 + 88 + 8 ~ ______________________ = ______

22 + 59 + 92 + 77 + 47 ~ ______________________ = ______

109 + 35 + 66 + 83 + 21 ~ ______________________ = ______

3. *What is the value of these Roman Numerals?*

VI ______ XXX ______ XVII ______ XL ______

LX ______ LXXXIV ______ CCLXIX ______ CCXCIV ______

4. 76 − 45 = ______

5. 359 − 169 = ______

6. 548 − 369 = ______

7. 704 − 388 = ______

8. 3,052 − 1,564 = ______

9. *A deposit of $755 was put down on an air fare costing $2,540. How much still had to be paid?* ______

10. *A house was bought for $137,559 and sold for $185,220. How much was made on the sale of the house?* ______

11. *The following scores were recorded for tests throughout the semester – 72%, 88%, 69%, 90%, 100%, 78%, 84%, and 85%. What was the average score?* ______

12. *Make a tally to show the color of cars in your street.*

Car Color	Tally	Total

13. *Change these 24-hour times to 12-hour times.*

1330 ______ 1520 ______ 2400 ______ 2010 ______ 1645 ______

14. *How many....*

minutes in $3\frac{1}{2}$ *hours?* ______

hours in 4 days? ______

seconds in $5\frac{1}{2}$ *min?* ______

days in 5 weeks? ______

hours in $3\frac{1}{2}$ *days?* ______

weeks in $4\frac{1}{2}$ *years?* ______

ANSWERS

Special Note: Where measuring is required for answers, the answers provided are only a guide. The process of printing and photocopying can alter the size of original artwork.

Assignment 1 - Addition

1. 12 14 16 16
   12 18 17 23
   21 19 37 21
   29 30 30 35
2. 48 88 78 99 100 80
   366 449 481 685 864 485
3. 277
4. 1,668
5. $1,062
6. 101,708
7. $46,495
8. 22,403
9. $2,956
10. 230
11. 6,537,050

Assignment 2 - Subtraction

1. 17 2. 41 3. 214 4. 106 5. 300
6. 393
7. 17 8. 13 9. 103 10. 174
11. 279 12. 1,667
13. $19.10
14. 24,455
15. 129
16. $3,725
17. 4,163
18. 6.5 L
19. 4,773
20. $26,250
21. 4.9 m 22. 3.3 m 23. 6.8 m
24. 8.2 m

Assignment 3 - Multiplication

1. 46 82 159 144 330
   336
2. 244 268 580 1,230 1,276
   2,625
3. 1,050 640 950 3,120 3,350
   2,970
4. 1,488 8,880 9,856 30,000
   27,995 5,900
5. $15 6. $90
7. 259 km 8. 9,984 L
9. 432 10. 12,448
11. 752 12. $171.60
13. 5,685
14.

6	12	18	24	30	36
5	10	15	20	25	30
4	8	12	16	20	24
3	6	9	12	15	18
2	4	6	8	10	12
x	2	3	4	5	6

11	121	110	99	88	77
10	110	100	90	80	70
9	99	90	81	72	63
8	88	80	72	64	56
7	77	70	63	56	49
x	11	10	9	8	7

Assignment 4 - Division

1. 6 8 7 24 32
   22 21 31 31 62
   31 31
2. $6 101
3. $11 $131
4. 14 14 23 18 29
   17 142 81 141 91
   141 131
5. 81 6. 45
7. 150 8. 15
9. 103r3 102r3 105r1
   267r2 74r1 122r5
   230r2 143 79r3
   65r8 291r2 34r9
10. 138
11. $98
12.

[1] 2	1	■	[2] 3	2	[3] 6
3	■	[4] 2	8	■	4
[5] 1	[6] 2	0	■	[7] 3	■
■	7	■	[8] 1	2	[9] 1
[10] 7	■	[11] 5	7	■	4
[12] 1	4	0	■	[13] 4	9

Assignment 5 - Mixed Problems

A- 9, 4, 40, 13, 24, 3, 0 ,10, 10, 3, 28, 2, 14, 5, 30, 15, 10, 7, 16, 5, 12, 5, 40, 2
B- 18, 17, 5, 10, 40, 26, 4, 0, 3, 19, 1, 4, 36, 16, 11, 14, 33, 23, 4, 6, 35, 101, 0, 10
C- 1, 24, 27, 6, 4, 18, 30, 18, 2, 24, 21, 10, 2, 18, 29, 12, 10,9, 8, 13, 8, 0, 21, 15
D- 8, 22, 36, 3, 10 ,12, 25, 6, 1, 40, 24, 2, 20, 25, 24, 9, 19, 20, 0, 10, 30, 50, 20, 8
E- 32, 3, 27, 11, 56, 6, 18, 17, 50, 2, 19,18, 24, 3, 22, 1, 44, 24, 18, 30, 4,15, 22, 10
F- 49, 11, 1, 40, 18, 5, 4, 19, 22, 8, 12, 15, 36, 12, 10, 21, 35, 4, 2, 29, 64, 19, 5, 28

Assignment 6 - Rounding

1. 20 50 10 20 100
   30 80 90 90 450
2. 100 600 300 400 700
   900 400 700 500 2,900
3. 2,000 4,000 4,000 7,000
   9,000 6,000 12,000 24,000
4. 2 min 16 min 10 min
   59 min 16 min 31 min
   46 min 14 min
5. 250 320 730
6. 70
   130 190
   70 20
   20 30
   800 800
   3,000 4,000
7. 180 410
   800 800
   190 14,480
   1,050 13,000
8. 178 409
   996 468
   189 14,485
   1,055 12,997

Assignment 7 - Place Value

1. 32; 2,006; 107; 655; 86,005; 321; 1,457; 20,573
2.

	Place Value	Meaning
5<u>7</u>	ones	7
<u>2</u>7	tens	20
<u>1</u>59	hundreds	100
30<u>8</u>	ones	8
<u>5</u>91	hundreds	500
1,3<u>6</u>7	tens	60
<u>2</u>,907	thousands	2,000
5,00<u>1</u>	ones	1

3. 8 80 80
   800 8,000 8,000
   8 80,000 800
4. 97,531 86,420
   77,665 98,610
5. 3,689 24,799
   2,356 14,789
6. 2 5 3 4
   36 4 9 1
   15 6 7 9
   5 0 0 0
   84 0 3 7
   70 5 0 0

Assignment 8 - Number Sentences

1. 15 8 8 22
   25 8 8 Answer may vary.
2. 16 5 7
   80 38 3
3. > < <
   < < >
   < > =
4. False True
   False False
   True True
5. 10; 12; 14; 1,632; 64; 128; 256
   15; 20; 2,510; 20; 30; 40
   6; 9; 12; 1,814; 17; 20; 23
6. Answers may vary.

ANSWERS

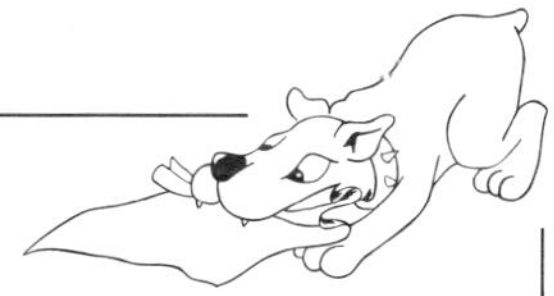

7.

9						
8						
7						
6						
5						
+	10	11	12	13	14	

50					
40					
30					
20					
10					
+	11	22	33	44	55

20					
19					
18					
17					
16					
−	5	7	9	12	14

100					
80					
60					
40					
20					
−	4	7	12	15	19

Assignment 9 - Multiplication

A	B	C	D	E
6	32	22	25	18
10	12	18	14	9
18	50	16	12	25
20	8	40	24	33
15	24	24	44	16
20	24	40	27	20
50	36	22	28	40
36	28	27	15	14
24	42	20	36	4
35	30	30	45	21
12	55	12	21	48
10	8	48	18	36
42	35	60	6	30
32	30	48	54	33
24	0	40	60	36
55	49	66	45	72
0	44	0	60	56
48	63	66	77	64
80	60	88	0	63
84	72	70	100	88
110	81	110	96	72
77	99	80	110	108
96	90	121	120	84
99	132	108	120	12
144	54	70	72	56

Assignment 10 - Decimals

1. Answers may vary.
2. (a) 1.0, 1.90, 2.4, 2.8, 3.05, 3.50
   (b) 4.00, 4.08, 4.88, 8.08, 8.48, 8.88
   (c) 12.05, 12.5, 12.55, 21.5, 25.1, 25.5
   (d) 32.0, 32.02, 32.09, 32.32, 32.90, 32.99
3. < < < <
   < < < >
   < > > >
4. 3 9 12 46
   23 26 37 59
   5 9 5 29
5. 2.87 m, 3.04 m, 3.56 m, 3.84 m, 3.94 m, 3.96 m
6. (a) 252.2
   (b) 145.65
   (c) 25.64
   (d) 117.73
   (e) 203.856
7. $14.00
8. 205.5kg
9. $766.23
10. $166.05, $61.50

Assignment 11 - Special Numbers

1. Teacher check.
2. 50
3. 50
4. 2 6 8 15
   30 16 19 60
   65 125 166 179
5. IV VII XII XXI
   XVII XXXIV XXXIX LII
   LXXXVI XCIX CXXVI CCCLXVI
6. Sean, Grant, Robert, Mark, Stuart, John, Michael
7. 2, 3, 5, 7, 11, 13, 17, 19, 23, 29, 31, 37, 41, 43, 47
8. Teacher check.

Assignment 12 - Money 1

1. $1.10 $1.95 $2.20
   $5.45 $10.10 $25.40
2. 75¢ $1.20
   $1.30 $2.50
   $1.05 $2.30
3. $4.00 $2.50 5¢ $3.75
4. $15 $18 $10 $4
   $9 $1 $4.50 $9.20
   $6.50 $11.80 $15.75 $18.10
5. (a) Answer may vary.
   (b) $11.40
   (c) $1.30
   (d) garlic seafood
   (e) $39.10
6. Teacher check.

Assignment 13 - Money 2

1. (a) $22
   (b) $44.50
   (c) $65
   (d) $7
   (e) $31.50
2. Answers may vary.
3. $46.80
4. $356
5. $64
6. $199,750
7. $111,425
8. $74,000
9. $274,500

Assignment 14 - Fractions

1. (a) $\frac{1}{2}$ (b) $\frac{3}{4}$
   (c) $\frac{3}{8}$ (d) $\frac{2}{8}$
   (e) $\frac{4}{9}$ (f) $\frac{6}{14}$
2. Teacher check.
3. 2 4 10 20
   2 5 10
   5
   2 4
   4 2 8 4
   5 10 3 3
4. $\frac{2}{4}$ $\frac{3}{4}$ 1 $\frac{4}{5}$
5. $\frac{2}{4}$ $\frac{4}{7}$ $\frac{3}{5}$ $\frac{3}{9}$
6. > > > >
7. $\frac{1}{5}$ $\frac{2}{5}$ $\frac{3}{5}$ $\frac{4}{5}$ $\frac{5}{5}$
   $\frac{1}{8}$ $\frac{2}{8}$ $\frac{3}{8}$ $\frac{4}{8}$ $\frac{5}{8}$ $\frac{6}{8}$ $\frac{7}{8}$
   $\frac{1}{10}$ $\frac{2}{10}$ $\frac{4}{10}$ $\frac{5}{10}$ $\frac{7}{10}$ $\frac{8}{10}$ $\frac{9}{10}$
   $\frac{1}{8}$ $\frac{1}{4}$ $\frac{1}{2}$ $\frac{3}{4}$ $\frac{4}{4}$

Assignment 15 - Problem Solving

1. (a) 940 km
   (b) 401 km
   (c) Safe Cove
   (d) 65 km
   (e) 562 km
2. $188.95
3. $359.40
4. 641 cm, 12 cm
5. 522%, 87%
6.

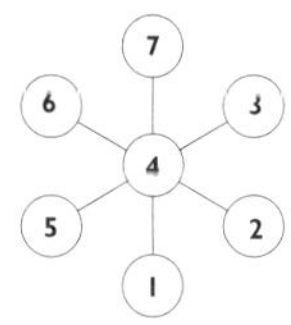

Assignment 16 - Calculators

1. $27.14 $44.45
   $55.30 $24.27
2. 121,502
3. 5,789 2,222 9,850 2,139
   14,248 5,752
4. 12,288
5. 5 8
   302 13
6. 55 87 91
   110 49 77
   501 4.5 25.5
7. 47,013
8. 1,793,083.4

Assignment 17 - Codes

1. America is a great place to live.
   Which state do you live in?
2.

12	7	8
5	9	13
10	11	6

6	11	4
5	7	9
10	3	8

12	27	6
9	15	21
24	3	18

10	3	11
9	8	7
5	13	6

23	19	18
15	20	25
22	21	17

3. 3, 2, 74, 3, 7, 1 5, 6, 3 7, 9, 3
   8, 4, 19, 3, 3, 6 3, 6 3, 7, 4
4. Answers may vary.

Assignment 18 - Measuring

1. 10 100 1,000
2. Answers may vary.
3. Teacher check.
4. 300 cm 200 cm 450 cm
   1,000 cm 260 cm 350 cm
   485 cm 860 cm 695 cm
   208 cm 770 cm 917 cm
5. 2 m 8 m 10 m
   3.58 m 0.49 m 4.4 m
6. 50 cm 200 cm
   5 m 405 cm
   10 cm 0.8 m
7. 12 cm
8. (a) 12.23 m (b) 44 cm (c) 24 cm

Assignment 19 - Perimeter

1. (a) 15.5 cm (b) 20.5 cm(c) 15 cm
   (Answers may vary due to printing and photocopying.)
2.

length	width	perimeter
4 cm	2 cm	**12 cm**
10 mm	5 mm	**30 mm**
25 m	10 m	**70 m**
5.5 m	3 m	**17 m**
12.5 cm	6.5 cm	**38 cm**

3. Answer may vary.
4. Answer may vary.
5. 80 cm
6. 11.8 cm
7. Answer may vary.

Assignment 20 - Area

1. (a) 8 (b) 9 (c) 6 (d) 12 (e) 14 (f) 17
   (g) 3 (h) 12 (i) 8 (j) 17
2. A = 16 cm^2 A = 18 cm^2
   A = 36 cm^2 A = 36 cm^2
3. Answer may vary.
4. Answer may vary.

Assignment 21 - Mass

1. 1,000
2. 1,000
3.

GRAMS	KILOGRAMS	TONS
toothpick	computer	truck
biro	adult	
empty lunch box	3 bricks	
library book	4 gold bars	
small doll		
2 marbles		
exercise book		

4. 1 kg 3 kg 2.250 kg
   4 kg 8.426 kg 3.415 kg
   2.005 kg 2.098 kg 3.456 kg
5. 2,000 g 4,000 g 3,500 g
   3,800 g 5,250 g 10,000 g
   12,400 g 16,800 g 13,550 g
   5,800 g 11,000 g 6,800 g
6. Answer may vary.
7. Answer may vary.
8. 500 g 750 g
   1.5 kg 500 g
   1,000 g 302.1 g
9. 176 kg
10. Answer may vary.

Assignment 22 - Tallies

1. (a) A (b) C (c) 76
2. Answer may vary.
3. Answer may vary.

Assignment 23 - Graphs

1. (a) 30 (b) ginger ale (c) cola
   (d) 6 (e) 5 (f) 27
   (g)lemon and raspberry (h) 12
2. Teacher check.
3. Teacher check.

Assignment 24 - Time 1

1. Teacher check.
2. 1.00 p.m. 3.30 p.m. 8.15 p.m.
   2.45 p.m. 4.10 p.m. 7.20 p.m.
   10.40 p.m. 6.00 p.m. 6.10 a.m.
   1.01 a.m. 12.00 a.m. 12.00 p.m.
3. 24, 60, 30, 60
4. 48 36
   240 300
   150 14
5. Answer may vary.
6. Answers may vary.

Assignment 25 - Time 2

1. (a)2 h 30 min (b) 3 h 15 min
   (c) 4 h 15 min (d) 4 h 55min
   (e) 8 h 5 min (f) 2 h 20 min
   (g) 2 h 20 min (h) 3 h 35 min
   (i) 5 h 30 min (j) 13 h 50 min
2. Answers may vary.
3. 3 h 45 min
4. 1,155
5. 2 h 40 min

Assignment 26 - Calendars

1. 12
2. 365 366
3. 31 28 31 30 31
4. Answer may vary.
5. Answer may vary.
6. January 2
7. Answer may vary.
8. Four
9. July 7
10. (a) Friday (b) November 12
    (c)19 (d) December 26
    (e) 30 (f) Three (g) December 7
11.

Winter	Spring	Summer	Fall
December	March	June	September
January	April	July	October
February	May	August	November

Assignment 27 - Shapes

1. (a) square pyramid (b) cylinder
   (c) rectangular prism
   (d) cube (e) cone
2.

SHAPE	VERTICES	FACES	EDGES
CUBE	8	6	12
CONE	1	2	1
CYLINDER	0	3	2
SQUARE PYRAMID	5	5	8
RECTANGULAR PRISM	8	6	12

3. Answers may vary.
4. Rectangular prism
5. Teacher check.
6. Answers may vary.
7. Answers may vary.

Assignment 28 - Tessellations

Teacher check.

Assignment 29 - Reducing and Enlarging

Teacher check.

Assignment 30 - Angles

1. ABC =118° DEF =133° GHI =158°
   JKL=135° MNO = 45° PQR =15°
2. Answers may vary.
3. ABC = 143° DEF = 40° GHI = 111°
   JKL = 78° MNO = 90°

Assignment 31 - Networks

1. (a) Yes (b) Yes (c) No (d) No (e) Yes
   (f) Yes (g) No (h) No (i) Yes (j) Yes
   (k) Yes

Assignment 32 - Word Search

S	H	A	P	E	S	E	M	A	S	E	L	G	N	A
E	F	R	A	C	T	I	O	N	S	S	S	A	M	C
L	E	L	Z	Z	U	P	N	M	E	E	S	E	E	A
B	R	E	D	U	C	E	E	U	N	T	E	N	A	L
A	A	S	E	D	O	C	Y	L	T	S	L	L	S	E
T	R	P	E	R	I	M	E	T	E	R	G	A	U	N
N	E	T	W	O	R	K	S	I	N	I	N	R	R	D
I	A	S	O	L	V	E	D	P	C	G	A	G	I	A
N	M	I	L	S	E	N	I	L	E	I	I	E	N	R
E	C	U	B	E	V	E	N	I	S	D	R	L	G	S
T	S	I	Z	E	I	M	O	C	L	A	T	N	E	M
Y	N	E	T	C	R	A	I	A	D	E	E	M	I	T
N	E	T	I	A	N	E	T	T	N	C	N	X	M	N
O	E	E	U	L	A	V	C	I	U	I	E	N	E	O
I	R	I	A	P	L	R	A	O	O	D	V	U	L	I
T	R	A	C	E	S	A	R	N	R	N	S	M	B	S
I	T	H	G	I	A	R	T	S	U	M	S	B	O	I
D	I	V	I	D	E	S	B	U	I	L	D	E	R	V
D	S	R	O	T	A	L	U	C	L	A	C	R	P	I
A	G	R	A	P	H	S	S	L	A	M	I	C	E	D

Mystery Word - millimeters

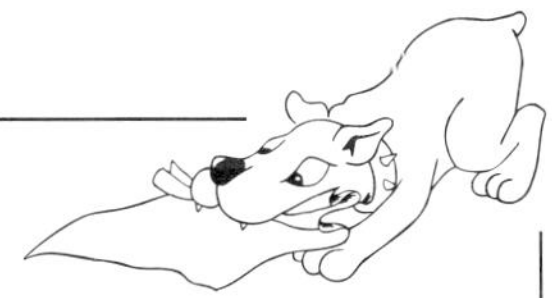

Assignment 33 - Review 1

1. Teacher check.
2. Answer may vary.
3. 854
4. 1,613
5. 8,554
6. 38.443
7. 29,234
8. A = 142° B = 55° C = 130° D = 38°
   E = 78°
9. A = 115° B = 85° C = 125°
   D = 120° E = 95° F = 10°
   G = 150° H = 20° I = 155°
   J = 108° K = 20° L = 78°
10. 25 cm²
11. 80 cm²
12. 3.0, 3.1, 3.5, 3.6, 3.7, 3.8, 3.9
    15.0, 15.1, 15.3, 15.4, 15.6, 15.8
13. Answer may vary. 14. December 2
15. February 6
16.

32	49	60	54
70	25	32	36
88	81	48	36

Assignment 34 - Review 2

1. 8
2. $11
3. $6.50
4. 54
5. $\frac{3}{5}$ $\frac{15}{20}$ $\frac{7}{8}$
6. Teacher check.
7.

11	21	1	8
45	64	3	28

8. 68; 156; 2,909; 20,003; 173,412;
   1,000,000
9. ones tens hundreds
   thousands ten thousands
   hundred thousands millions
10. 1,000; 1,000
11. 1 kg 5 kg 2.5 kg
    10.5 kg
12. 3,000 g 4,500 g 7,900 g
    10,250 g
13. 400 cm 350 cm 1,000 cm
    550 cm
14. 3 m 6 m 5.5 m
    9.85 cm

Assignment 35 - Review 3

1. 84
2. 260
3. 180
4. 804
5. 3,300
6. 11,400
7. 180
8. $42
9. Teacher check.
10. true true
    false false
    true true
11. 7, 9, 11, 13 20, 30, 40, 60
    21, 28, 35, 42 6, 5, 7, 6, 8
12. $1.25 $1.50
    $1.80 $4.80
    $6.40 $3.50
13.

$45	$35	$25
$19	$1	$29.50
$18.25	$7.20	$35.55
$43.75	$23.70	$15.85

14. $2,590
15. $10,356.54
16. (a) 12.3 cm (b) 20.7 cm
    (c) 19.2 cm (Answers may vary due to printing and photocopying.)
17. 55.88 cm

Assignment 36 - Review 4

1. 40; 90; 120; 300; 360
   900; 1,210; 2,340; 2,560; 410
2. 370; 360; 240; 300; 320
3. 6, 30, 17, 40
   60, 84, 269, 294
4. 31
5. 190
6. 179
7. 316
8. 1,488
9. $1,785
10. $47,661
11. 83.25%
12. Teacher check.
13. 1.30 p.m., 3.20 p.m., 12.00 a.m.,
    8.10 p.m., 4.45p.m.
14. 210 min, 96h, 330s, 35 days, 84h,
    234 weeks